Technological Risk

Technological Risk

H. W. Lewis

W·W·Norton & Company · New York · London

Printed in the United States of America.

First published as a Norton paperback 1992

Library of Congress Cataloging-in-Publication Data

Lewis, H. W. (Harold Warren)
Technological risk/H. W. Lewis.
    p.   cm.
1. Technology—Risk assessment.   I. Title.
        T174.5.L48   1990
    363.1—dc20   90–30460

ISBN 0–393–30829–4

W. W. Norton & Company, Inc., 500 Fifth Avenue, New York, N.Y. 10110
W. W. Norton & Company Ltd, 10 Coptic Street, London WC1A 1PU

3 4 5 6 7 8 9 0

Contents

Part 3   Coda

Acknowledgments

Over the last fifteen years so many people have helped me learn about risk that it would be impossible to name them. It was Pief Panofsky who, as president of the American Physical Society in 1974, tempted me into this career deviation by asking me to chair the Society's study of reactor safety. It was Robert Oppenheimer who, as my major professor and friend many years ago, set such an extraordinary model for the breadth of knowledge necessary to take on real applications of physics. Theoretical physics in the years after World War II would have been far poorer had he not been around.

Among those who have read and criticized parts of the manuscript, I owe special thanks to Bill Press, who read it all; Bill Anders and Charlie Miller, who read the chapter on air safety; the group at the General Motors Research Laboratories—Bob Frosch, Dick Schwing, and Len Evans—who read the chapter on highway safety; Joe Keller and Alice Whittemore, who read the chapters on toxic and carcinogenic chemicals; and Dick Wilson, who read the chapter on fossil fuels. What errors remain are, of course, mine.

My editor, Ed Barber, actually read the manuscript in detail, and demonstrated how much I still have to learn about writing. Somewhere he has learned to criticize without giving offense, and I am in his debt.

Finally, many acknowledgments end with thanks to the author's family for having endured the distractions and strange hours that go with writing (especially for people with other real jobs), but I had always thought those thanks were simply courtesies. They aren't.

Introduction

We are obsessed with risk, especially in the novel forms brought to us by science and technology. Risk has become a major political and social issue, provoking widespread uneasiness about scientific progress; demagogues thrive in such an atmosphere.

Yet the risk is real, as are the benefits. We live our lives surrounded by the miracles of modern chemistry, yet we are preoccupied with chemical contamination. We use clean electricity from nuclear power plants, yet fear the prospect of a nuclear accident. Our lives have been dramatically extended by vaccines against many of the former scourges of mankind, yet the few cases in which a vaccine has done more harm than good are widely publicized, and concerned parents often refuse to immunize their children against known diseases. Commercial aircraft speed increasing numbers of us across the country in a matter of hours, ten times more safely than we could drive the same distance, yet fear of flying lingers. Fluoridation of our drinking water is making tooth decay a thing of the past, yet fear of chemicals deprives much of the nation of the benefits of fluoridated water.

We are both beholden to technology for enrichment of our lives and suspicious of the associated risks—especially when they are unfamiliar. It is an uneasy cohabitation. It would be good to know the price, including risk, that we have to pay for the benefits, but both risk and benefits are hard to estimate, even for dedicated experts. In all those mentioned above there are genuine benefits and genuine risks. Because they are not easy to balance we often do the job badly, accepting unnecessary risk in some more familiar forms, while grossly exaggerating it in others. Most people would probably agree that there is no point in exposing ourselves to a risk for which we get nothing in return, and probably even agree that we ought to accept great risks if the stakes are high. The history of the human race would be dreary indeed if none of our forebears had ever been willing to accept risk in return for potential achievement. Risk is part of the price we pay for growth, as nearly all parents know.

Just as there are trivialities that scare us witless, so are there real risks that don't bother us. Many agonize over the possibility that, in a thousand years or so, high-level nuclear wastes in geologic storage may leak (the current EPA standard requires no leakage for ten thousand years), while paying almost no attention to the fact that, every year, about a thousand Americans are accidentally electrocuted. No mass demonstrations oppose electricity, no rock concerts ring with speeches clamoring for lower distribution voltages, even though reduced distribution voltages would make an accidental shock less harmful. (Most of the world uses higher domestic distribution voltages than our 110 volts, with consequent economic gains.) Lightning kills about a hundred Americans each year, yet the sale of lightning rods is far from a sure road to riches. The most egregious example of a risk only beginning to be taken as seriously as it ought to be is smoking, which kills nearly four hundred thousand Americans each year, well over a thousand a day, often unpleasantly. One out of five American deaths is from smoking. To be sure, in the more than twenty years since the Surgeon General issued the famous report that certified that smoking causes lung cancer, smoking rates in the United States have gradually turned around, peaking in 1971, but we are dragging our feet. Incomprehensibly, many doctors still smoke, along with nearly 30 percent of the adult population of the United States. If we spent as much per untimely death caused by smoking as we do on coal mine safety, there would be no money left in the United States for any other purpose—it would require the entire gross national product. We even cough up over $30 billion per year just to buy cigarettes. That's still less than the estimated $100 billion we spend on illegal narcotics, about which no more will be said.

In fact, cigarette smoking, though declining in the United States, is growing as a world problem. In the Third World it is growing at a faster rate than the population, and in Africa almost twice as fast. Perhaps the same social forces are at work that are present here, where smoking is more prevalent among the less well educated than among the more educated, among blue-collar workers than among white-collar workers, among blacks than among whites, and so on.

This book is meant to improve public appreciation of the difficulties of risk assessment and management, and to contribute to public understanding of the issues. It is aimed at intelligent readers, not specialists. Though there are many forms of risk, the book is confined to the risks connected with progress in science and technology. The other threats to life are important—they dominate the mortality (death) and morbidity (sickness) statistics—but they are not the subject here.

Many of the risk-related issues, like the value of life or the problems of the regulation of risk, don't depend much on the specific form of technological risk under consideration. These matters are dealt with in Part I. On the other hand, there is no better way to learn than by example, so Part II is a potpourri of specific cases, treated in greater detail, and grouped in families with common features. Part II is meant to satisfy the craving for real issues that has been built up in Part I. The collection is far from complete, but each example is chosen to be of current interest, and to illustrate some point made earlier. Finally we have to recognize the unfortunate fact that no deep appreciation of risk can be developed, particularly with respect to the likelihood of rare events, without some minimal knowledge of statistics and probability. Part III contains the necessary lore, aimed at an inquisitive reader with a grounding in high-school mathematics. No one should be ashamed of using numbers. Part III also contains a kind of summing up.

Something must be said about timeliness. It is tempting to seek timeliness, but research and legislative and regulatory change make it impossible to be entirely up to date. Further, there is an argument against going too far in the direction of currency, since newly discovered information that has not stood the test of time may later turn out to have been wrong—that happens routinely in science. To be up to date is to be quickly out of date. This book will be reasonably, but not entirely, up to date.

Concern about technological risk is a function of both location and time—it is a relative newcomer on the scene, most visible among the more developed nations. The point of departure here will be unabashedly American, both through the examples chosen and the discussion of the regulatory forces at work, though variants of the issues and the examples can be found throughout the world.

We should also be honest about the significance of risk compared to the other threats and challenges facing mankind. Though technological risk is important, it is far from the *most* important subject. When doom comes to our species it will not be from trace chemicals in our air or water, nor from a nuclear accident. As T. S. Eliot foresaw, the end is less likely to come with a bang than with a whimper. We should reflect occasionally on the implications of the fact that the earth is grossly overpopulated and becoming more so, especially in the Third World. The First World is not insulated from the economic and population problems of the Second and Third Worlds; we're all in this together. The term "spaceship Earth" has been coined to describe this state of affairs, and the time scale for solving the population problem is one or two generations. It cannot go on this way, and the die is already cast.

Part 1

Generalities

1
The Risks of Life

Fear and risk are different creatures. What some of us fear most—poisons in our drinking water, radiation in our air, pesticides on our food—pose hardly any real risk, while some we fear least—driving, drinking, and smoking—kill many hundreds of thousands each year.

And risk isn't all bad. Personal development is impossible without risk—how would anyone learn to ride a bicycle? On a larger scale, evolution would be impossible without the risks and challenges that strengthen species. For better or worse, we would never have become the creatures we are without the risks to which our forebears responded.

This chapter is devoted to collecting the known facts about the threats to life in the United States, to set the framework for the rest of the book. Later we'll concentrate on technological risk.

For now let's think of risk as the chance of death before our allotted time. Of course death is inevitable for us mortals, so it may seem odd to use a certainty as a measure of risk. Alternatives are sometimes used, like the number of years or days or minutes by which a life is unnaturally shortened. By the first measure, death ten years too soon is treated as no worse than death five years too soon; by the second, it is counted as twice as bad. Some contend that the early years are more valuable than the later years, while some (like Robert Browning) argue otherwise. Chapter 7 will deal more deeply with the contentious subject of the value of life. When a famous comedian reached the age of fifty, and people began to speak of him as middle-aged, he is alleged to have said that he would feel better about it if he knew a few more hundred-year-olds. In fact, about 1 percent of us will live to be a hundred, and more than 80 percent of those are likely to be female. For the moment, the chance of premature death is an adequate definition of risk.

The average expectation of life in the United States has been rising steadily for many decades. An American born in 1920 could expect a life span of fifty-four years—females a year more than males—while by 1985 the life expectancy had risen to seventy-five. The worldwide average is about sixty. The premium for being female in America is now seven years, and decreasing. Our life expectancy has therefore gone up over the last sixty-five years by about four months for each year. If this were to go on for the next seventy-five years, the average American would live to be a hundred, and the recent crisis over the solvency of the Social Security fund would appear in retrospect to have been child's play. The increase in longevity is mainly a result of the conquest of some real scourges of the young, through medical advances and public health measures. Infant mortality in the United States has been cut in half in the last twenty years, a dramatic improvement that still leaves us behind twenty other industrialized countries. There has been a smaller improvement at the more advanced ages, where the average remaining life at age fifty has only been going up at a rate closer to a month per year, and is now about twenty-eight years. To get to age fifty, the most important step is to get to age one; 95 percent of those who make it that far can expect to live the remaining forty-nine years.

The difference between young and old in terms of improvement in survival prospects is even more evident if we go still further back in time. When the Constitution took effect in 1789, over two hundred years ago, life span data were kept in Massachusetts. At that time, and in that place, the expectation of life at birth was about thirty-five years, compared to seventy-five today. By contrast, the remaining life expectancy at age sixty was fifteen years in those days, and has now climbed to twenty. For old folks, the prospect of a long life ahead is not so much better now than it was then. Of course, people who made it to sixty in the late eighteenth century must have been pretty tough.

When we finally do pass on to the ultimate resting place, what is the cause? The table on the next page lists the leading causes of death in 1985, while the graph displays the age distribution of the major contributors. The items in parentheses in the table are the major subcategories of the previous item, so that, for example, nearly half of all fatal accidents are associated with motor vehicles.

The table reflects a kind of snapshot in time, and is not a prediction of the future, if only because our population mix is not stable. There are nearly twice as many births as deaths in the United States, so we are a young crowd, and will stay so for some time to come. That affects

Cause	Number
Cardiovascular diseases	978,000
(Heart disease)	(771,000)
Malignancies	462,000
(Lung, respiratory, etc.)	(127,000)
Accidents	93,500
(Motor vehicle)	(45,900)
Pulmonary diseases, chronic	75,000
Pneumonia	68,000
Diabetes	37,000
Suicide	29,500
Liver disease	27,000
Homicide (including police)	20,000
Other	. . .
Total	2,086,000

Major causes of death in the United States in 1985

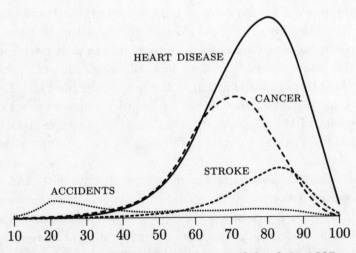

Age dependence of various causes of death in 1985

the table by giving more weight to diseases that primarily afflict the young. For instance, the figure below the table shows the peak age for cancer deaths at seventy-two, while that for heart disease is eighty, so the latter will increase in importance as the population ages. (People don't think of cancer as a comparatively young persons' disease, but it is.) The accidental deaths are represented by the lowest curve on the figure, are dominated by traffic accidents, and peak at age twenty. (Suicide is the second-largest killer of Americans between fifteen and thirty-five.) As subcategories of non-auto accidents, though we haven't shown a chart for this, drownings and poisonings are most important in the twenties and thirties, while falls, the second most important cause of accidental death, are most important for those over sixty, peaking in the eighties. All of these age-dependent effects will affect the relative importance of the various causes of death in the future. So will medical research, which will conquer current diseases, and biological dynamics, which will create new ones. In 1900 the leading cause of death in the United States was tuberculosis, a disease that is now responsible for less than one death in a thousand. And AIDS was unknown.

Note also that the graph is a display of the total number of deaths of each type in 1985, and is not the death *rate*. The fact that all the curves show a peak followed by lower values at the more advanced ages doesn't mean that we develop an immunity to death as we get older, only that there are fewer of us left around to die. Only about 5 percent of our population is over the age of seventy-five, a proportion that would increase to 8 percent if we had a stable population with the current mortality rates. Yet the ratios shown on the graph tell us a great deal about relative risks. Thus, at age sixty cancer and heart disease kill about the same number of us, while by age ninety heart disease kills five times as many as cancer.

Not only do most of these causes of death depend on age, they are also specific to the United States at this time. The suicide rate in Austria is twice as high as ours, and that for Italy half as high. The death rate from heart disease is a third as high in France, and a fifth as high in Japan. The death rate for stomach cancer is almost four times as high for the Japanese. The table is truly a cut in both space and time, and not the permanent condition of the human race.

The obvious messages are that the real killers in our society are circulatory diseases and cancer (responsible for 47 percent and 22 percent of all deaths respectively in 1985), and that technology has contributed very little to the fatality rate. To be sure, handguns are used in about half of the murders in the United States, and one in a hundred of us is doomed (statistically) to be murdered, but handgun technology hasn't changed as much in recent decades as has handgun proliferation. Handguns don't represent technological risk in the way that possible chemical and nuclear accidents do. On the other hand, it is fair to ask whether technology plays a role in the increase in cancer rates in recent decades, or whether the aging of the population is responsible. We owe technology for much of that longevity. Even there, the age-adjusted fatality rate has been decreasing for some time for nearly all types of cancer. There are outstanding exceptions for lung and respiratory cancer, where the undoubted cause is smoking. Only fanatics and tobacco merchants continue to deny this. Just as an example, the death rate (per hundred thousand) for such cancers, for women aged fifty-five to sixty-four, has increased by more than a factor of five in twenty-five years, from seventeen in 1960 to ninety-four in 1985. These are women who started smoking after World War II, when it became fashionable to do so, and the risk was unclear. This is the price for that token of emancipation—lung cancer is now a bigger killer than breast cancer for such women. Of course, women still have a long way to go before they catch up with men in smoking mortality. The fraction of males who smoke has decreased about 35 percent in the last twenty-five years; for females the decrease is much smaller, about 15 percent for the same period. Among high-school seniors, more females smoke than males. For people without a high-school diploma, smoking has hardly decreased at all. There is a social aspect to smoking.

Technology contributes most to our mortality table through motor vehicle accidents, since they would certainly not have occurred if the automobile had not been invented. Yet travel by horse over comparable distances would have been riskier, so even the automobile probably saves lives. Of course, this cannot be proved, because people would never have traveled as much on horses or on foot. Even now, we are more likely to be killed by a car (per mile traveled) when walking than when we are driving. It has been observed that if we really want to save pedestrians' lives, we should put them in cars.

Other widely feared technological hazards, like nuclear power, simply don't appear in the table because they cause so few fatalities. In the case of nuclear power, the fear is of a catastrophic event, so the fact that no one was killed by nuclear power last year is not especially persuasive to the fearful. There is a whole class of risks which appear as the remote probability of a calamitous event, and the rational treatment of such risks—to neither overplay nor underplay them—is not trivial. We do rightly worry about a major nuclear accident, of which there has been at this writing just the one at Three Mile Island in this country, which killed no one, and the one at Chernobyl in the Soviet Union, which has killed thirty-one so far; we do worry about large airplane crashes, of which there are typically one or two per year; we do worry about devastating earthquakes, which occur less than once per generation in the United States; we do worry about the collapse of a large dam, and so on. Some even worry about visits from alien civilizations, less likely than any of the above. The challenge is to do our worrying constructively, without paralyzing our civilization.

In addition to all these rare but familiar eventualities, there are the truly apocalyptic technology-based risks, whose damage lies in the distant future. The burning of coal and oil has combined with massive deforestation of the earth, the latter to accommodate the spread of population (there are more of us, we need room to live, and we have an inclination to eat), to cause a steady and easily measurable increase in the carbon dioxide content of the atmosphere. No one doubts that this will affect the climate, but we are not sure by how much, or when, or what the effects of the climate change will really be. It may or may not be true, for example, that worldwide flooding of coastal regions, from the melting of the Antarctic and Greenland ice caps, is in the cards. Though some melting would inevitably accompany a worldwide warming trend, we cannot yet predict just how much.

The same is true of health effects from fluorocarbon-induced ozone depletion of the atmosphere, or the poorly quantified climatic effects of nuclear war (the so-called nuclear winter), or a number of other grand-scale risks. All are real, posing long-term threats to humanity, but all are sufficiently remote to give us a little time to get our act together. Whether we have the will and wisdom to do so is another matter.

2

The Measurement of Risk

Webster's Unabridged *New International Dictionary* (the revered Second Edition) says that risk is "hazard; danger; peril; exposure to loss, injury, disadvantage, or destruction." It distinguishes risk from hazard by suggesting that a risk is more often voluntary, a hazard the product of chance. Hazard itself is an old game, of which craps as we know it is a simplified form, and is defined in the (equally revered) *Oxford English Dictionary* as "a game at dice, in which the chances are complicated by a number of arbitrary rules." These definitions contain the essentials of risk, combining the idea of loss with that of chance or probability. The latter is crucial, since the inevitable may be unpleasant, but it lacks the element of chance and is not risk. Death and taxes are presumed to be inevitable. Only efforts to evade the latter are risky, though the risk is accepted by many in return for potential gain.

Mathematicians define probability as a number between zero and one (a fraction, if you like) that measures the chance that something will happen. A probability of one means an event is a sure thing, while a probability of zero means it is impossible. A probability of 1/2, or 0.5, means it is likely to happen about half the time, a tossup, or even money. In everyday life, people rarely speak of probability in this sense, but they do speak of odds, and they even gamble. To say that the odds are two to one that a given team will win a football game means that the chances of winning are two out of three, and of losing, one out of three. The probability of winning is thus two-thirds or 0.667, a number between zero and one. Odds and probability are always related in this way, so a probability of 0.01 can be thought of as ninety-nine to one odds against. Most people are more comfortable with the idea of odds than with probabilities, presumably because of the experience most of us have had with gambling. In fact the members of our society with the deepest practical knowledge of probability may well be the professional gamblers. They know far more than the amateur gamblers, who squander their substance on illusions like "winning streaks" at dice and lucky numbers at horse races. According to a recent poll, half of the American people believe in lucky numbers.

There are four broad categories of risk, covering most cases.

1. The familiar high risks, which exact a large toll, and on which we have good information. Driving and hang-gliding are reasonable examples.
2. Risks of low probability, whose consequences are so large that they must be taken seriously. An example might be a large earthquake.
3. This category might be considered an extension of the second—events whose probability is so very low that they have never happened at all, yet, but whose

prospective consequences are so awful that they deserve attention. An example might be a major destructive change in the climate, as a consequence of atmospheric pollution.

4. Finally a collection of substantial risks which, though real enough, are hard to evaluate because they show up as increases in naturally occurring hazards. Examples might be any of the cancers caused in part by environmental contaminants, where the additional incidence is hard to separate from the "natural" rate.

An example of the first category—familiar risk—might be the chance of demise in an automobile accident, while driving from Los Angeles to San Francisco and return. So many have done this, and so many have regrettably perished along the way, that the odds of survival are known. Average statistics show that there are about three trillion passenger-miles accumulated in the United States each year, in passenger cars, with about 45,000 fatalities due to all forms of motor vehicle travel. Only(!) about 25,000 of these are actually occupants of passenger cars—we'll see all the categories in Chapter 13—so the chance of getting killed is just about one in a hundred million miles of travel, for an average occupant.

The trip is about four hundred miles each way, so the risk of untimely death is about a chance in a hundred thousand. About one in forty thousand Americans dies each day, so the incremental risk of death from the trip is equivalent to the normal likelihood of dying in somewhat less than half a day. The risk seems acceptable, and there are mighty few who would make the risk estimate before setting out on the trip. Further, each driver is inclined to consider himself (or herself) so skillful that nothing can conceivably happen. Besides, when we have done something successfully for a long time we become complacent. It's a truism that no living driver has had the direct experience of being killed in an automobile accident, and lack of firsthand knowledge affects our attitudes.

We worked this out in some detail, just to show how this kind of well-known risk is calculated when there is long experience and a wealth of information. There was no need to even use the nationwide statistics, since that particular trip is taken so often that specific data are available. The calculation could even have been subdivided according to young drivers or old drivers, male drivers or female drivers, freeway or scenic route—all the relevant information is available. These common risks are easy to analyze, because we have the data.

Estimates of the *consequences* are more difficult, involving highly personal questions of value, which most people prefer to avoid. For example, a typical accident insurance policy available at airports doesn't only insure a passenger's life. It pays the same benefit for loss of life as for the loss of any two items from a list consisting of one's hands, feet, and eyes, but only half the benefit for the loss of one item from the list. While gruesome, that implies that someone has made a judgment about the value of these body parts to the average airline passenger, and has equated eyes to feet, etc. (People haven't always thought that such tradeoffs were appropriate. In Exodus, the Bible says, "life for life, eye for eye, tooth for tooth, hand for hand, foot for foot ....")

It is not easy to assign value to such losses. When dealing with tangibles, like property, it is useful to speak of replacement cost, or repair cost, or some such measure, but that kind of estimate tends to lose its cogency when applied to things that cannot be replaced or repaired. (It is relatively easy to replace people, but impossible to replace specific people.) A passionate debate therefore rages about the value of life, or even the value to be assigned to those irreplaceable items that contribute to the quality of life. What is the value of a beautiful sunset, or of an ancient redwood grove, or of Yosemite, or of Aunt Martha?

In any case, for the first category of risk—the common and familiar threats—the probabilities are available from statistical analysis of a surfeit of data, and assessment of the consequences is difficult only because matters of subjective judgment are often involved. Is a foot *really* equal to a hand or an eye? What is it worth to be able to walk across a street safely—more than eight thousand pedestrians get killed each year. Should there be traffic lights and crossing guards on every corner in each town with a population over ten thousand? A thousand? A hundred?

This question and the sunset question lead us to an important and widely misunderstood point. There are those who argue fervently that there is no limit to the resources we ought to expend to preserve life, and they take a dim view of risk analysts who try somehow to make an assessment of both sides of the equation. To do so does require putting a monetary value on a life, a limb, and on a beautiful sunset, so that we can judge how much to spend to preserve them. Money is, after all, our medium of exchange; its purpose is precisely to make it possible to trade in the value of things without bartering the things themselves. Yet analysts who try to set a value on life can expect a torrent of abuse for just making the effort. (When the Ford Motor Company did such an analysis for the threat of fire from collisions involving the Pinto fuel tanks, and decided that the value of the potential lives saved equated to an expenditure of about $11 per car, they were excoriated. It was not that the number they chose for the value of life—$200,000—was too small, it was that they had dared to set any value at all.) The standard expression used by opponents of risk/benefit analysis is that one is "comparing apples with oranges," which they believe impossible.

Actually, that unveils the flaw in the argument. Given a fruit bowl containing both apples and oranges, few of us will have much trouble making the choice. Given a shopping expedition to the supermarket, only the unusually indecisive shopper will have a problem. One could even test consumer response to different relative prices of apples and oranges, just to see which they like best, and indeed fruit merchants and supermarkets do that, so they can sell both commodities. The price represents a combination of relative cost to the seller and relative desirability to the buyer, as it should. Of course these value judgments are subjective; we are, after all, *not* machines. We make choices, and are often hard-pressed to defend them on logical grounds. So what?

The same point can be made about the need for traffic lights. Our society doesn't actually behave as if the lives of pedestrians were priceless. We allocate a certain level of resources to our democratic government, and that government allocates a certain fraction of those resources to traffic lights, after which we all accept the residual risk. Though the world is full of people who pontificate about the preciousness of life (and of course we all feel that way about the lives of those near and dear to us), we simply don't act accordingly.

The second category of risk is harder. Here the risk is real, but the probability so low that experience provides little guidance, yet the consequences are potentially so high that we are rightly concerned. Consider the chance of a major earthquake in Southern California, where this author lives. We know from occasionally exciting experience that we live in an earthquake-prone area, riddled with faults (cracks in the earth), and are smugly superior when visitors from the East are terrified by our routine small quakes. Major devastating quakes like the 1906 San Francisco disaster don't happen often, but are inevitable. Earthquakes are caused by the steady movement of the great tectonic plates of which the crust of the earth is made, and cannot be prevented. We can limit the damage they cause by building more resilient buildings, by research which can lead to improved understanding and perhaps even reliable warning, by improving emergency response capability, by public education, and so forth. All of those measures involve costs, and our willingness to incur these costs ought to depend on the probability that the event will occur in some reasonable time. We get our estimates of the probability from limited experience, and from theoretical understanding, neither of which is a precise guide. As a practical matter, we simply wait.

But we do try to expand our knowledge. We do as much research on the underlying causes as the country is willing to support, so that the sparse data can be used most effectively to make predictions. We have models, albeit imperfect, of earthquakes and floods and other natural phenomena, and reasonable estimates of their frequency. Since it has now been nearly a century since the last massive earthquake on the San Andreas fault in California, and since the internal stresses along the fault can be measured and are increasing, it is truly only a matter of time until there is a major event. (When this was first written, the California earthquake of 1989 had not yet occurred. It was small compared to the 1906 earthquake, and we are still waiting for the "big one.") We don't know where it will occur on the fault, when it will occur, or how much havoc it will wreak. Both the probabilities and the consequences are uncertain, and under those conditions we tend to do very little. At the University of California in Los Angeles a faculty committee estimated a few years ago that a number of the older buildings might not withstand a major earthquake, so there could be substantial loss of life if the event were to occur while those buildings were occupied by students. Yet there is no sense of urgency, and there are other demands on the scarce resources. The same can be said about the potential for dam collapse in the area.

The third major category of risk—things that have never happened at all, yet *could* happen—is even more remote. If they also don't do much harm we should forget about them, but if the results could be devastating, we ought to try to protect ourselves where we can. Examples abound, some technical and some not so technical. The consequences of a major nuclear war among the great powers would surely be so dreadful that we and the Soviets spend a substantial part of our gross national products on both equipment and activities designed to reduce the probability. It is unpopular in this country to try to do something about the consequences (like build shelters), because that is misinterpreted by some as accepting the inevitability of nuclear war.

This is the category that suffers most from misunderstanding of what is implied by a low probability. A low probability means only that; it does not mean that an event will never happen or can somehow be avoided. Acceptance of the fact that small probability doesn't mean zero probability is the beginning of understanding. In the case of nuclear war, one has to believe that the probability is precisely zero to insist that efforts to mitigate the potential consequences are misguided. Even though we all hope and work to avoid such a catastrophe, it is wishful thinking and a kind of arrogance to assume that success is guaranteed, and that we need not contemplate failure. According to the Rogers Commission report, that kind of misunderstanding of probability contributed to the loss of the space shuttle Challenger in 1986.

There are many more examples of events of high consequence, and either low (but not zero) probability or remoteness in time. The impact of a large meteorite on a city could kill millions, but since little can be done about it we don't even try. The steady increase in the atmospheric concentration of carbon dioxide has already been mentioned, with its potential to cause worldwide climatic catastrophe. Nuclear war itself.

This low-probability high-consequence category is in one way the most interesting of the four, since the estimates of *both* consequences and probabilities must be based entirely on theory. The probabilities come from probabilistic risk assessments (more later) with inevitable uncertainties, and the estimates are apt to generate disagreement and confusion. There is nothing intrinsically wrong with disagreement—democracies are supposed to thrive on it—and uncertainty itself is a legitimate and honorable feature of any scientific enterprise. Yet legitimate uncertainty provides an opening through which demagogues and technical charlatans can get into the act, and can exert disproportionate and ultimately destructive influence on public decision making. It is a source of great anguish to this author that so much of our public policy on technological risk is determined by lawyers acting as if they were technical experts, and by show business personalities trading on their commercial success in playing fictional roles. Both are out of their element.

There are many examples for this category, all different. We have not yet experienced a major commercial nuclear accident in this country, yet it is inevitable. Such are the laws of probability—if the probability is not zero, the event is ordained, given enough time. We need to know the probability, the likely consequences, and the best ways to reduce the probability *and* the consequences. We spend a great deal of money on regulation of the nuclear industry (the budget for the Nuclear Regulatory Commission alone is over $300 million annually), but continue to lack any agreed sense of "how safe is safe enough." Therefore we have no way to know when we've done enough, and should point our efforts elsewhere. The nuclear energy community is responsible for many of the most impressive advances in risk analysis in the last fifteen years, and also for much of the support for research in seismology (an earthquake can damage a nuclear power plant), but the probability of a major accident remains remarkably uncertain. That opens the door to the mischievous, who exploit uncertainty.

The fourth and last category—known risks that are increased slightly by technology—is often the most frustrating. It includes a whole menagerie of threats for which both probabilities and consequences are elusive. Not because the effects are unfamiliar, but because they are so familiar, and the extra damage is such a small increment in our already imperfect world. The classic examples are the health threats posed by low levels of natural or commercial chemicals, and the effects of low levels of radiation. The fear of contamination (for health, not æsthetic reasons) sometimes approaches the classic symptoms of a phobia, yet the threat is real, and we are well advised to do our best to understand it. Our problem in assessing the probability of harm from these contaminants is that the effects are so small, despite the clamor they often generate. We can make a few general comments here, but each is a separate case, and we'll go through some examples in Part II.

There is no doubt that certain chemicals in the environment, or in our food or drinking water, can cause cancer in people at some concentration, with exposure over some period of time. In no case do we know a magic exposure level below which a chemical is "safe" and above which it is dangerous. The important question in understanding chemical carcinogenesis is to determine the cancer induction rate for different exposures and exposure periods, so that public policy can be geared to reducing the risk to an acceptable level. It cannot be reduced to zero, especially since many of the worst offenders are in the natural environment. (Aflatoxins, found in peanuts, are among the most powerful carcinogens known, but who would think of banning peanuts at baseball games?)

The difficulty is that, since cancer is such a common disease (22 percent of us now die of some form of cancer), it is statistically almost impossible to determine just which cancer cases are due to what cause. Sometimes particular forms of cancer can be associated with identifiable exposures—that's how we know, beyond doubt, that smoking is by far the major cause of lung cancer in the United States—but we can't measure cancer induction directly at very low exposure levels. The chemical effects, if any, are swamped by the more than 400,000 "normal" cancer deaths in the population each year. (Some experts believe that oxygen is a culprit in the normal occurrence of cancer. It would be interesting to see an effort to purge the atmosphere of oxygen, or to make it illegal to breathe.)

It is the same with the effects of low levels of radiation. We know that large doses of radiation can cause cancer (or, paradoxically, sometimes cure it), but again have no way to learn whether that is true of the low doses that are associated with such things as the normal cosmic-ray background or competently delivered medical and dental x-rays. At low doses the effects are simply too small to be measured, and it may be that the lowest doses are harmless. But we don't know. If we wish to administer our society in such a way that we are not exposed to excessive risk, yet also don't overreact to small or negligible threats, we are in a pickle. Chapter 15 is devoted to this.

Finally, we have to deal with what the experts call risk aversion, relevant to all four categories. Up to now we haven't made any distinction in importance between the probability of an event and the consequences of that event, though risks that have *both* low probability and low consequences have been ignored—that is just common sense. It remains to ask whether a better measure of risk can be formed from some combination of the probability of an event and its consequences. Which matters more?

People running an insurance company find this easy. They multiply the probability of the event by the amount of the potential loss, call the product the expectation of loss, and use that as a final measure of risk. If an event has a chance in a thousand of happening, but would cause a loss of a million dollars, the expected loss is a thousand dollars. A non-profit insurance company, if one could be found, would charge that as a premium. This works because the fictitious insurance company might have a thousand such policies to cover, which would lead it to cover one loss on the average, using up the million dollars it had collected in premiums. Multiplying the probability by the value of the loss is good economics, and most reputable insurance companies function that way, adding to the premiums enough for overhead, profits, advertising, salesmen's salaries and commissions, and other perceived necessities. They would behave this way no matter how large the policy, provided the potential loss doesn't have the potential to bankrupt the company. (That would change the rules, as is known by many gamblers who have bet everything on the last roll of the dice.)

But people are not insurance companies, and there is a school of thought that holds that larger losses have to be considered worse than would be implied by the value of the loss, and therefore have to be held to even lower probabilities than the multiplication procedure would suggest. In that way of thinking, one large loss is worse than two smaller ones, even when they add up to the same amount of damage. An event that could destroy ten thousand homes would represent a higher risk than one that could only destroy a thousand, even if the probability were ten times smaller. This reasoning is what drives some companies to require that two high corporate executives not travel on the same airplane. (They often share a limousine to the airport, which can be riskier.) Though the probability that at least one will be killed is twice as high if they travel separately (either airplane can crash), the probability that both will be killed is much lower, and it's more important to avoid that catastrophe. Many people seem to think that way.

Much of the argument against nuclear power has risk aversion as its basis—though the probability of a major accident is extremely low, the consequences could be so severe that the technology is deemed unacceptable.

In the rest of the book we'll measure risk by the product of the probability and the value of the loss, the way of the ideal insurance company. Even some major insurance companies occasionally lose track of the familiar rules. In 1971 Lloyd's of London was asked to write an insurance policy to protect the Cutty Sark liquor company against the possibility that someone might catch the Loch Ness Monster that year. Cutty Sark had offered a prize, and was suddenly (and inexplicably) concerned that it might have to pay. Lloyd's had no way to judge the probability of capture, especially since the monster may not exist, but wrote the policy anyway, charging an outrageous premium that bore no relation to any reasonable probability of loss. Cutty Sark paid, apparently without haggling. Nobody made any pretense of calculating the product of probability and consequences. (It was pure profit for Lloyd's—the monster wasn't caught. Surprise.)

3

The Perception of Risk

We in the affluent societies are preoccupied with safety, while risk is recognized as a normal condition of existence by the less fortunate. Somehow the strange idea that the world owes us a risk-free life is a relatively localized and recent phenomenon, pretty much confined to the Western industrialized world. Even in the United States it tends to have regional emphasis—more in the West than in the Midwest, more in the North than in the South, and so on. Perception is very personal, and generalizations can be treacherous.

One perception of risk has a long history among primitive peoples, still flourishes in some segments of society, makes risk easier to accept, and appears in many of our insurance policies—denial of its existence. Insurance policies often refer to acts of God, implying that what happens is ordained by a higher authority, and is not just chance. It is a form of fatalism. Those who firmly believe that misfortunes are inflicted on us from above ought to go no further, because rational risk management won't help in

that case. What is actually done to manage our exposure
to risk depends, of course, on whether we think it threatening, and whether we think we can do something about
it. The first of these is the subject of this chapter.

How, then, do people recognize risk? What criteria
are used to determine whether a risk is acceptable? What
makes us fear some threats while ignoring others?

In particular, what is meant by the word "risk" when
dealing with the ultimate fear, loss of life? Up to this
point it has been the likelihood of untimely death (when
is death not untimely?), but there is more to risk than
that. Insurers openly, and most people subconsciously,
carry some sort of assessment of the relative values of life
(both shortening and loss) and of various deformities or injuries. We mentioned airport insurance policies in the last
chapter. We all have different values, based on such factors
as early experiences, education, circumstances, personality, etc. Chapter 1 began with the observation that fear
and risk are two different creatures, yet an individual's
response to risk depends very much on the nature of his
fears, rational or irrational. Consider some examples.

Looks, to some of the more fortunate among us, can
assume such a central role as to be more important than
life itself. In California, disfigurement through involuntary tattooing has been found by the courts to be equivalent to mayhem, a term encompassing loss of a limb or of
some other vital part. Yet voluntary tattooing continues
to thrive, so at least some feel otherwise. Of course it depends on the nature of the tattoo. Even in California it is
illegal to tattoo a minor, voluntarily or involuntarily.

Some will jeopardize their lives to protect property,
and most will do so to protect loved ones from injury.
Many still cling to the old values, and will risk their lives

to protect their country, while all too many will do so to promote their religion. The dictum (still adhered to, but now considered sexist) that in an emergency one ought to sacrifice the men to protect the women and children has its roots in a distant era, long forgotten, when there was a shortage of people. The world does not now suffer from such a problem.

We will deal with loss of life through two principal measures: the chance of untimely death itself and some version of the so-called YPLL, years of potential life lost. The latter is often used among risk analysts, most often counting only the years before age sixty-five as *really* lost. This gives expression to the view (shared by few senior citizens) that a risk imposed on old folks is somehow less damaging to society than if it threatened the young. It is sometimes assumed that the value of a person's remaining productive life is proportional to his or her remaining life expectancy, or to the remaining pre-retirement years. A person with twenty years remaining is then considered twice as valuable as one with only ten. Such a procedure actually overvalues the economic worth of the twilight years. In cold impersonal economic terms, the very old and the very young consume more than they produce, while those in the middle years support themselves and the others. Over a lifetime, we more or less break even, and neither productivity nor personal attachments are directly related to the number of years left. Yet medical statistics are commonly expressed in terms of YPLL, and the objective of medical procedures is life prolongation—not salvation, which is the responsibility of other specialists. In the version of YPLL in which only the lost years before age sixty-five are counted, the lives of those over that age are regarded as valueless—hardly a defensible position.

We'll use the term risk loosely, most often just in terms of life lost.

Even the term "untimely" poses problems when used to describe death. The Centers for Disease Control use the term premature to describe a death that occurs before the age of sixty-five, and publish death rates in all three forms: YPLL (before age sixty-five), premature deaths (again before age sixty-five), and crude total mortality rates. The table in Chapter 1 used crude mortality, and heart disease and cancer led the list of causes of death. But heart disease afflicts older people more than does cancer, as was illustrated through the graph in Chapter 1, so a comparable table for premature deaths will have these two causes nearly equal. (The graph shows that they run neck and neck up to about age sixty, at which point cancer levels off, before starting to decline in the years after seventy-two, while heart disease continues to increase. By age ninety, the vast majority of all deaths are attributable to either heart disease or stroke.) If we take the next step, and list causes of death in order of YPLL before sixty-five, both of these take a back seat to unintentional injuries, or accidents, which are the leading cause of death among the young. In terms of YPLL the four big ones are accidents, cancer, heart diseases, and suicide/homicide, in that order. At age twenty-one accidents are by far the leading cause of death, and three-quarters of those are motor-vehicle accidents. There are even regional and local differences—in terms of YPLL the District of Columbia is nearly twice as dangerous as the worst state in the union. Can one ask about the value of life without specifying *whose* life?

The elements most frequently mentioned as affecting an individual's perception of risk are:

1. Is the risk voluntarily assumed, or imposed by outside forces? Smokers often rely on this for part of the rationale for continuing to smoke. "It's my choice!" Risk is even easier to accept for a member of a group sharing the same situation, since mutual reinforcement can suppress any lingering doubts about the wisdom of the course. In the military world, this is known as *esprit de corps.*

2. Is the risk familiar or unfamiliar? Ghost story writers, carnivals, and demagogues exploit fear of the unknown. This is one reason why low-probability risks often seem worse than those of high probability—they are bound to be less familiar. The least familiar are, of course, those that have never shown their faces.

3. Does the risk lead to immediate harm, or is the day of reckoning far in the future? With apologies to Omar Khayyám, heed not the rumble of a distant drum.

4. How is the risk expressed? People are extremely vulnerable to verbal cues—this is known to risk analysts as the framing question.

The first of these requires little discussion. People are quite willing to assume risk, sometimes just for fun and sometimes as part of the job. When they do, it is often with little clear idea of the magnitude of the risk, provided they believe they control their own destinies. This stood out clearly in the mid-1970s, through the widespread resistance to compulsory seat belts. During the short-lived reign of automatic seat-belt interlocks (the pesky things that kept you from starting your car unless you were belted in), it was fashionable to disconnect them as soon as a new car was delivered. The dealer would offer the service as a freebie. In the end, the interlocks were so frequently disconnected that they had to be abandoned.

Both air bags and mandatory seat-belt laws have the same problem. Though the open argument about these unquestionably life-saving measures is about their cost and effectiveness (exaggerated by enthusiasts in the heat of the debate), the submerged issue of coercion *vs.* voluntarism is at least as important. Who likes the do-gooders who arrogate to themselves the right to protect us from ourselves? We didn't ask them to.

We've already mentioned the second item—familiarity of the risk. People exaggerate the risk in the unaccustomed. More people are still afraid of flying than of driving, though the fatality rate for commercial aviation is about one fatality per billion passenger-miles, with automobiles ten times worse. Xenophobia, the fear of other countries and cultures, has sometimes been justified by experience, but unfamiliarity plays an important role. Americans are notorious for their suspicion of those who don't speak English. The most exaggerated current fear is probably that associated with the storage of high-level nuclear waste, with deep roots in the unfamiliarity of radiation. Despite the essentially unanimous view of informed scientists and engineers that the risk is grossly overrated, the fear remains. Yet the radiation level in this room, as this is being typed, is higher (because of trapped indoor radon and its products) than it would be directly on top of a nuclear waste repository. Anyone who really fears nuclear radiation ought not to write a book indoors.

The third factor, timing of the consequences, is both more interesting and more subtle. Some risks pose immediate threats to life and limb—driving, mountain climbing, walking under ladders, entering a tiger's cage, and the like. For these we could in principle work out the chance of catastrophe and judge whether the activities carry, on balance, compensatory rewards. Bad luck carries consequences both timely and unmistakable. You take the chance, and pay the piper if you lose.

On the other hand, a person exposed to a dreadful disease (like AIDS or leprosy, or smoking-induced emphysema or lung cancer) may or may not contract the disease as a result of the exposure, and the consequences may be a long time coming. AIDS typically takes over ten years, while the consequences of smoking or radiation exposure or exposure to chemical carcinogens can be delayed twenty or thirty years, or even more.

At the far end of the time scale, there are risks whose damage lies in the distant future, affecting future generations. Major climatic change from the continued burning of coal and other fossil fuels (the greenhouse effect) is generally estimated to be twenty to a hundred years away (though it could come sooner), while the opponents of a nuclear waste repository speak animatedly of what might happen in a thousand to ten thousand years. (To see the futility of looking that far ahead, imagine the people of ten thousand years ago planning for our welfare. That would take us nearly back to Cro-Magnon. With the best of intentions they couldn't have known how to help.)

In this author's direct experience, the record for distant vision is held by a former governor of California, who worried about the welfare of the people who will live (we hope) a hundred thousand years in the future. If instead we think back a hundred thousand years, about the time *Homo sapiens* first appeared on earth, we might ask what those early ancestors could have done for us, other than survive. In the grand scheme of things, and assuming we wish well for the human race, survival is the single most important duty we have toward our descendants. Everything else, including the quality of life, comes after that. If we don't survive, they won't exist. It is pretentious to suppose that they will share our sense of values, or that we can predict their needs.

How then should we deal with risks that threaten future loss? There must be some outer limit for concern, a time beyond which there is just no point in fretting, whatever our sense of responsibility. The self-destruction of the sun is currently scheduled for a few billion years from now, but it makes no sense to prepare for it. We rightly pay less attention to risks whose day of reckoning is far far away. For most people the span of concern extends through their own lifetimes and those of their children—for some, not nearly that far. Of course we should prepare for and tend to the future, out of a decent sense of social responsibility, but should do so with humility and perspective.

Economists and bankers have no problem dealing with future contingencies—they do it every day—and the mechanism is discounting, as with an annuity or compound interest. The latter is more familiar to most people. With an annuity we make small payments now in the expectation of greater rewards (a comfortable retirement) in the future, while with compound interest we voluntarily give up the use of our money, so that it can build to larger amounts in the future. In both cases the value of something in the present is deemed to be higher than if it were deferred to a later time. (Bird in the hand, and all that.) No law of nature asserts that the future is less relevant than the present, but people have behaved for many centuries as if it were. Experience even provides a guide for the rate.

Given a choice between a calamity this year—say a broken leg—or one in ten years, few will decide to get it over with. Anything can happen between now and then, and even the Devil can be cheated of his due, if we are to believe Stephen Vincent Benét. Conversely, given a choice between a gift of a million dollars this year, or an inheritance of the same amount in ten years, it would take some kind of nut to opt for the latter. Deep in our hearts, we do know that we should discount the future, though the discounting rate depends very much on psychological factors like our sense of security. Banks and investors make such discounting a practice, and it is possible to learn about our collective wisdom from their experience. Judging from the rates used for savings accounts, investment returns, and similar trades of present wealth for future benefits, the rate at which we have historically discounted the future seems to be between 5 and 10 percent per year (over inflation, of course).

This practice goes both ways. We are willing to give up present good for even greater rewards later, which is called investment, and we insist on paying less now for something to be delivered in the future, which is called discounting. They are in principle the same. It is only one step from there to the concept called insurance, which means that the policy holder doesn't wait for loss, but pays premiums now as a hedge against a future cost. A trust fund is the same as investment; resources are squirreled away (with interest) for future withdrawal. If the future withdrawal is to deal with a future contingency, and is made available at the time of the contingency, it is called insurance, and the investment is called a premium.

All of these financial shenanigans are based upon the same concept—anything destined for the future is worth less now, whether it is good or bad. As was said earlier, this doesn't have the status of a law of nature, but is simply an observation about millennia of human behavior. (Incidentally, none of this is to be confused with inflation, which simply devalues the unit—money—in which things of value are measured. Allowance for inflation is an additional consideration when using money as the medium for future planning.)

To put it more explicitly, suppose the discount or interest rate is 10 percent, and an investor wants to have $10,000 at the end of a year. It would be necessary to invest $9,090.91 now, so that the accrued interest of $909.09 would bring the total to the needed $10,000 after the year. If the investor could wait two years, the needed investment would be $8,264.46, and so forth, so that the accumulated compound interest and principal would add to the $10,000 after the proper time. Ten years would reduce the needed investment to $3,855.43, and twenty years to $1,486.43, which is beginning to look like a bargain. Less than 73¢ will do it in a hundred years. (This is not investment advice.)

So the systematic way to deal with future risk is to treat it as if it were a business loss, and to determine its current value by discounting it at a rate of 5 to 10 percent per year. In 1972, the United States Office of Management and Budget (OMB) ordered all federal agencies to use a discount rate of 10 percent when calculating the costs and benefits of any intended actions. That was OMB's assessment of the proper level at that time. The order is still in effect, but widely ignored.

Most will agree that discounting is the logical procedure for dealing with the future (though some deny vigorously that it is relevant to questions of human life and health), but there is considerable controversy about the "right" rate. Reasonable people may differ, reflecting their

different views of the value difference between the future
and the present. Once the principle is accepted, however,
there are dramatic consequences for decision making on
risks; one will always devote more effort and resources
to dealing with immediate problems. The future will be
treated fairly, neither overrated nor underrated—both errors are common.

This issue of future loss will return in Part II, but a
simple grand example can illustrate the point. The gross
world product in 1986 (total of goods and services produced anywhere on earth) was recently estimated at something like $15 trillion, $15,000,000,000,000. Suppose we
knew of an impending disaster (say, due to carbon dioxide buildup) which threatened to reduce this by one-third,
a loss of $5 trillion, and a major setback for the human
race, but the event was not due for two hundred years.
Accepting the OMB recommendation of 10 percent as an
appropriate discount rate, we could now ask how much
should be spent today to avert such a tragedy. The answer
turns out to be about $25,000, as a one-shot investment.
That amount can either be spent now to avert the disaster, or invested in a savings account, so the $5 trillion will
be waiting in the bank when it is needed. (What bank,
you may well ask, but bank is just a figure of speech.)
A discount rate of 5 percent would have led to an answer near $300 million, and a reasonable number probably
lies somewhere between. Neither of these appears to be
particularly expensive as the cost of prophylaxis or cure
for a misfortune of that magnitude—a single airplane can
cost $300 million these days—and most of us would proba-
bly want to make the investment, out of an unquantifiable
sense of responsibility for future generations. We can think
in terms of two hundred years. If we were talking about a

thousand years into the future, that might be a different matter. The cost would be negligible, far less than one cent, but that is pretty far off, and we might simply not care.

This is all relevant to the perception of risk because, though all the above is economically beyond reproach, we don't all understand economics or compound interest, so we tend to waste current resources on distant threats. The billions being squandered on the quest for an absolutely safe nuclear waste repository provide a classic case, as we will see later. To make it worse, we also often err on the other side in this age of instant gratification, and both government and corporate leaders are reluctant to invest in the future. Very few of our major industries now support basic research, even in their own areas of interest. Marketing of today's products is deemed more important than the development of tomorrow's. It is a curious paradox that aversion of future harm seems more important than the promise of future benefit. That was not always true. Those who are unwilling to invest in the future haven't earned one.

The fourth important contributor to the perception of risk is the way the risk is portrayed—the framing question. It is easy to illustrate by asking people to make decisions on risk questions, when the same questions are alternately framed in terms of either benefits or losses. It turns out that they will usually take chances to minimize or avert a loss, risking a greater loss, but are more likely to go for a sure thing to secure a gain. Obviously, as all observers of gambling casinos know, there are variations among people, but these are the most common patterns.

One way to make the point consists of offering people a chance to flip a coin for a prize of $1,000, with no loss if they fail, but offer them a chance before the toss to suggest a settlement payoff. That would be a kind of settlement out of court. It's a clear winner, a possible gain with no chance of loss, but what is it worth? Since an even-money chance of winning $1,000 gives an average expectation of gain of $500, one would expect people to be willing to sell the opportunity for about that much. Yet tests show that they will settle for about $350, on the average. They want the sure thing, even if they don't get as much, and are willing to pay for it. This is the gain case.

If, on the other hand, the rules are changed just a bit, so the player is *given* $1,000, and asked to toss the coin to decide whether he has to give it back, his decision changes. It's of course the same thing, except that now he has the money in hand, and the coin flip is for a loss, not a gain. The probability is again that he will get to keep the money about half the time, so the expectation of gain is still $500 (the $1,000 in hand, minus the 50-50 chance of having to give it back). The big difference is that he already has the $1,000 and can savor it, so the gamble seems to be to avert loss, not to win. A mathematician should still offer to settle by paying back $500 to forego the toss, or, if he is consistent with the previous choice, he should be willing to pay $650 to assure the net gain of $350. But no, players still offer only about a $350 settlement, preferring to gamble everything on the chance of not losing anything. To a mathematician or statistician, or even to a professional gambler, this makes no sense at all. A professional gambler should lick his chops at the chance of meeting someone like this.

So people will generally gamble to avoid loss, but are conservative about potential gains. There they prefer to take the bird in hand. This assumes, of course, that they know the odds—very low-probability events, like lotteries and catastrophic accidents, are dominated by lack of information and understanding, as well as by superstition. In some tests, people who have just bought lottery tickets for one dollar have been unwilling to sell them back for two dollars. The bird in the hand is, in this case, no bird at all.

It is worth noticing that the very use of the term "risk" to define the subject of this book biases the issue. Had we used the term "safety," the psychological tone would have been different. This is recognized in the naming of federal agencies, all of which are devoted (at least in name) to assuring safety, and none to reducing risk. The Department of Defense, despite its name, is meant to wage war if necessary. The Department of Health and Human Services deals more with sickness than with health. The Department of Justice runs the FBI. The Department of Energy spends more on nuclear weapons than on energy. But the Centers for Disease Control, bless them, do exactly what the name implies. Still, euphemisms are the rule rather than the exception.

Sensitivity to the way in which questions are framed extends far beyond risk evaluation (where it helps to explain the routine exaggeration of low levels of risk) to other areas like amateur gambling and amateur (and some professional) investment strategies. Framing bias is unavoidable in connection with risk, since risk is rarely expressed in terms of how many people have escaped it. Imagine describing a day in the life of the airline industry by saying "Yesterday a million passengers rode the commercial airlines, the vast majority of whom were eventually deposited alive at some destination." Not much chance of seeing that on a billboard.

But what does our citizenry think of all this? About ten years ago an Oregon-based research organization asked four different groups of people—members of the League of Women Voters, college students, members of a business and professional club, and finally experts—to rank thirty reputedly risky activities in order of risk. Just comparing college students with experts, they found that the college students rated nuclear power public threat number one, while the experts put it in the twentieth spot. The experts rated motor vehicles number one, while the students rated it fifth (after nuclear power, handguns, smoking, and pesticides). And so forth. These discrepancies between fact and fancy carried across to the other groups, perceived risk often more dependent upon media coverage than on actual risk.

We mentioned in the Introduction a particularly egregious example of priority confusion, the decades-old question of fluoridation. By now, after a whole generation has grown up with drinking water to which a trace (about a part per million) of sodium fluoride has been added, many also using fluoridated toothpaste and mouthwashes, the results are in and are clear. Fully 50 percent of all children between the ages of five and seventeen have not had a single cavity in their permanent teeth. In the last decade alone, the number of children's cavities has gone down by about 40 percent. Further, the dire predictions for public health because fluorides are poisonous (true, in large enough doses, of many of the things we eat regularly, including fluorides) have not been borne out. Fluoridation has been as close to an absolute winner as can be found, nearly all benefit at minimal cost, with no detectable down side.

Yet the political pressures are such that, despite near

unanimity among experts, about 40 percent of the American people live in areas in which the water has less than optimal natural fluoride concentrations, and none is added. Three of the ten largest cities don't adjust their water. In the entire Los Angeles Basin only two communities fluoridate their water, and the City of Los Angeles is not one of them. In effect, the anti-technology forces have made life so uncomfortable for the City Council that action in the public interest is bad politics. There is no vocal political constituency for good teeth, so a politician whose principal concern is to assure reelection cannot afford to waste goodwill on teeth. Fortunately, many of the deprived children use fluoridated toothpastes.

As an example of the quality of the debate, a recent anti-fluoridation publication asserted that most cases of AIDS are in cities that fluoridate their drinking water. It could just as easily have said that they occur in cities with a public library. Equally true—equally irrelevant.

Fluorides represent just one example of a familiar situation in which the common good is ill served by the democratic process. The problem is exacerbated by the emergence of groups of persuasive people who specialize in technology-bashing and exploitation of fear, make their livings thereby, and have been embraced by large segments of the media as experts. The next chapter deals, all too briefly and kindly, with this national problem, and with the distortion of priorities it brings with it.

4

The Politics of Risk

Since risk is a subject on which passions run high, it is necessary to say something about the players, and common decency requires that it be done as fairly as possible. Still, some of the participants exert a destructive influence on rational decision making, with consequent damage to all of us, and it would be disingenuous to pretend otherwise. This is a very short chapter devoted to recognizing the existence, in the United States and in Western Europe, of substantial and effective political forces that are simply opposed to technology, and use their political strength almost entirely for obstructive purposes. The German Greens are the most open about their platform, while the American equivalents are less well focused. They evoke in many of us a genuine nostalgia for a simpler life, a reaction to the fact that our technological world is simply harder to understand. The sense that we have somehow lost control of our destinies is certainly depressing, and an anti-technology posture can strike a responsive chord. Opposition to change, especially technological change, is a full-time profession for many, and the term "activist" is now used proudly, as if activism were honorable in itself, regardless of what one is active for or, more often, against.

The problem is exacerbated by a frightening trend in our society. Just as our lives are becoming more complex, interactive, computer-oriented, and—let's face it—technological, our population is declining in its educational level. It is no secret that the average scores on the standardized Scholastic Aptitude and Achievement Tests have been decreasing for many years, though the rate of decline seems to have been arrested in the early 1980s, and the scores have crept up a bit since then. Still, they are not even close to what they were only twenty years ago (they are fifty points lower), and the tests have not changed that much in their level of difficulty. To boot, the curricula in our schools have been greatly softened in the last few decades, academic achievement has been de-emphasized as a proper objective of education, science and mathematics have given way to more "relevant" material, and grade inflation has disguised the decline in standards from students and parents alike. All this has been documented in any number of reputable and solid studies, and railed against by any number of reputable scholars. In contest after contest in which our students are pitted against their foreign equivalents, we turn up near the bottom of the heap. All parents support—even demand—better education for their children, as long as it is in the abstract, but far fewer if it impairs the children's enjoyment of life, or faces them with the unthinkable—failure if they don't perform.

A partial consequence of this denigration of learning is that the fraction of our population that believes in UFOs and reincarnation is mind-boggling, less than half of us know that the earth goes around the sun once a year, and it is an unending struggle to keep the teaching of evolution legal in the schools. Americans are about evenly divided on whether evolution or creationism is more correct. Half the American people believe in lucky numbers. Finally, as a direct consequence, it has been estimated that American industry spends as much on remedial mathematics education each year as is spent on direct mathematics education in elementary schools, high schools, and colleges.

Many of us are dependent on television pictures and sound bites for our information, and formulate our positions on peace, war, the environment, risk, and the economy from the one-dimensional heroes and villains we see for a few seconds on the evening news. Those who have learned to dispose of complex points in five seconds or less appear on the talk shows, and all the television news programs have learned that their audience share depends almost entirely on their entertainment value.

It is worse among the young, who were raised in a television atmosphere. According to a National Opinion Research Center poll regular newspaper readership has decreased from 75 percent of the population twenty years ago to 50 percent now, while in the 18–29 age group it has decreased from 60 percent to 29 percent. In the 30–34 group it has gone from 75 percent to 45 percent. It is mainly the old folks, the over-sixty brigade, who still read newspapers. It has become an article of faith, encouraged in the schools, that anyone's opinion is as good as anyone else's, whether or not informed.

You, dear Reader, are in a distinguished minority— you are actually reading a book. Not only a book, but a book, however unworthy, that is meant to leave you somewhat better informed than you were before you read it. If it doesn't succeed it is truly the author's fault, not yours.

Our very literacy as a nation is in danger. The current estimates of the Department of Education are that a full one-third of us, seventy million Americans over the age of seventeen, are either functionally illiterate or only barely literate. The number is estimated to be increasing by about two million per year, even though the great majority of our children now graduate from high school. (More than 75 percent of the adult population has been through high school, compared to 25 percent in 1940.) Even more of us are functionally innumerate (the numerical equivalent of illiterate). Not only is this a work force which must be accommodated in an increasingly complicated and demanding job market, but it is also a population with rights and privileges, including the ancestral right to the pursuit of happiness. Above all it is a substantial electorate with a decisive voice in the ways in which our country responds to the challenges of technology. We are a participatory democracy and it is everyone's country, not just the educated. At the moment these words were first written, in March of 1988, not one of the eleven remaining candidates for the presidential nomination in the two major parties had had any scientific or technical training, and that was no accident. (The solitary one who did was one of the first eliminated from the race.) This is the backdrop against which the anti-technology forces work. The problems described in the last few paragraphs go so far beyond the subject of this book that we will leave them in a moment, despite their transcendent importance.

However, it is hard to resist mention of an interesting and perhaps even relevant story. We are told that in 1968 a poll was taken of a "representative sample" of German women, asking them what profession they would prefer in an ideal husband. At the top of the list was nuclear physicist. (Since this author is a male physicist with nuclear pretensions, that is heady stuff. Physicists, especially theoretical physicists, don't get many chances at a good fantasy.) Alas, in 1979 a similar poll produced a very different answer. Nuclear physicists didn't appear anywhere on the list of the first twenty choices—top of the heap was forest ranger. While the Germans love their forests, that is quite a change of preference, whose significance is too painful to pursue.

Technological risk provides a testing ground for the ability of a democratic society to manage its affairs in such a way as to assure the common good. Garrett Hardin, in his 1968 essay *Tragedy of the Commons,* made the point that a society can be badly served if each member of that society makes decisions aimed at serving his own perceived self-interest. The composite of individuals acting in their own interests can easily translate into a situation in which not only the society but each individual comes out badly. His example was a common pasture, in which each individual can appear to benefit by adding to his herd, leading to overgrazing and destruction of the pasture. We control such problems by accepting the strictures of government, balancing each individual's freedom of choice against the demands of the general welfare. The underlying logic breaks down when the government itself reflects no more than the sum of the individual choices, as is the case in a participatory democracy, and breaks down even more when the participants are ill-informed. Even a minority can then do real damage.

There is, of course, a history. In the early nineteenth century in England, as part of the Industrial Revolution, automatic machinery was introduced into the textile industry. This had the inevitable result that skilled crafts-men (with now obsolete skills) found themselves unem-ployed, through no fault of their own. Their resentment was originally directed at the machinery itself, especially the stocking-frames (machines used in the production of knitted stockings, which made it possible for relatively unskilled workers to produce stockings at more than ten times the historic rate), and the five years beginning in 1811 saw widening convulsions of rioting and destruction of machinery, at first only textile machinery, but later a wider variety. The rioters were called Luddites because their leader assumed the name General Ludd, after a per-haps mythical Ned Ludd who was alleged to have destroyed stocking-frames in 1779, and therefore to have been some-what ahead of his time. The term "Luddite" has become opprobrious, and has come to mean anyone who is strongly opposed to machinery, or, by extension, to technology. It is quite appropriate here.

The Luddites finally generated a strong reaction in Parliament, which passed a law that made the destruction of stocking-frames a capital offense. After that a number of Luddites were hanged, and the movement crushed. The economic recovery that followed Napoleon's defeat at Wa-terloo, and the peace of 1815, probably also played a role in relieving the unemployment that fueled the riots.

An interesting sidelight of this miserable episode is the fact that Lord Byron's maiden speech in the House of Lords, at the age of twenty-three, was in opposition to the capital offense bill, and in defense of the Luddites. The distaste of poets for technology has a long history—read Walt Whitman's *Learn'd Astronomer* to get the flavor— but the cudgels have been taken up in modern times by the lawyers. There aren't enough poets to go around these days, and they can't be spared for this kind of work. On this one, however, technology won the next round. Byron's daughter Ada, Countess Lovelace, was one of the remarkable early figures in the history of computers, and a modern computer language is named after her. You can't keep a good gene down. Of course her mother was a mathematician.

Strangely, displaced workers do not form the core of the anti-technology movement today—it seems to be an upper-middle-class phenomenon. Such people are genuinely concerned that technology may be destroying the environment, and have presumably never seen the environment in other, less technically advanced, countries. (Mark Twain once said, "To be good is noble. To tell other people how to be good is even nobler, and much less trouble." With comparable relevance, H. L. Mencken once observed, "There's always an easy solution to every human problem—neat, plausible, and wrong.") It is a pity, since the environmental quality that is left to us does need vigorous and informed protection.

Up to this point, this chapter has probably (and not surprisingly) appeared one-sided, as if all the sins in the politicization of risk have been committed by the anti-technology elite. Unfortunately, that is not so. The counterpoint to the politics of ignorance is the politics of complacency, embraced by more than a few of the guardians and keepers of risky technology. The very low probabilities that go with some of the more unlikely threats to our health and welfare—nuclear accidents, nuclear war, the greenhouse effect (high probability in the future, but just beginning), etc.—can be misleading. The fact that the disaster has not yet come has led all too many to believe that it never will, and to relax. It is not easy to maintain vigilance when major accidents never occur, especially when the most strident prophets of doom are demonstrably uninformed, or even worse. Ignorance on the one side and complacency on the other (especially when combined with mutual contempt) are twin threats to the rational management of risk.

5

The Assessment of Risk

The assessment of a risk involves both probability and consequences, and we made a first survey in Chapter 2 of the means by which each of these is measured. This chapter will be devoted to a closer look at probabilities.

As was said at the beginning, an essential element of risk is randomness—the fact that we don't know exactly whether, when, and where the damage will occur. Of course there are also non-probabilistic ways to incur harm, but they are not risk. For someone who jumps off a tall cliff, there is little doubt that something bad will happen soon. (Despite the old joke about the man who fell off the roof of a skyscraper, and was heard to say as he passed a fourth-floor window, "So far, so good.") Doom yes, risk no. It is through the randomness that probability enters the fray, and it is crucial to the business of risk assessment.

Probability and statistics will get a more thorough treatment in Part III, where all of this will become more precise and complete, but there is one feature—probably

50

the most important one—that deserves top billing. That is the famous square-root-of-N rule, though no learned statistician would call it that. We'll simply explain the rule, and make it plausible by example. Anyone who understands the role of square roots in statistics is halfway home.

The rule—which is usually true, but has some exceptions—is simply that the uncontrollable fluctuations in the number of times a random event occurs are approximately equal to the square root of the expected number, give or take small factors. (Statisticians call this magic number the standard deviation.) Therefore, the amount of information one can get from event counting is fuzzy, or uncertain, by about the same amount. (These are not the same comment; in the first, one is going from information to observation, in the second from observation to information.)

For those who have forgotten, the square root of N, denoted by $\sqrt{N}$, is the number which, when multiplied by itself, makes N. Thus the square root of nine is three ($3 \times 3 = 9$), and the square root of twenty-five is five ($5 \times 5 = 25$). Most often, the square root is not an exact integer, but that's not important. For example, the square root of fifty is a number slightly larger than seven, because seven times seven is forty-nine, slightly less than fifty. In fact, the square root of fifty is approximately 7.071, a statement which is written $\sqrt{50} = 7.071 \ldots$, and one can add as many decimal places as one wishes, to make the answer more and more exact. In risk analysis, that is a waste of time. Consider an example.

Suppose our favorite baseball team has on it a powerful batter, who routinely bats .300 (a hit 30 percent of the time, a formidable average reached by twenty batters in the major leagues in 1988). Our man does it season after season, and is therefore indispensable to the club. Alas, in the last dozen or so games, during which he has been at bat fifty times, he has had only ten hits, so his batting average for that period is a paltry .200, normally considered beneath contempt even for poor batters. With a batting average so far below his norm, he may well be proclaimed by the sports columnists for the local newspapers to be in a "slump," and will probably come to the park early for the next few games, to take extra batting practice. The manager and coaches will meet with him to discuss what is wrong, and he may even be benched for a few days, to give him a rest. Does any of this make any sense? Not if we assume that each time at bat is an independent event (an assumption that may or may not be true), and look at the statistics.

During those fifty times at bat, he would "normally" have had fifteen hits, and no one would have been concerned. But the square root of fifteen is pretty close to four ($\sqrt{15} \approx 4$ is the way that would be written, where the wiggly equals sign means not quite equal, but close enough). No one should be surprised if, in any given stretch of fifty times at bat, he were to deviate by four hits in either direction. Ten hits, which is a deviation of five hits on the low side, is a bit out of range, but not enough over the line to be worrisome. In fact,

for this particular case, the probability that our
hitter gets ten or fewer hits in fifty times at bat,
when he is hitting normally, is 0.079, so the odds
against it happening as a random event, without
any real significance, are a bit worse than eleven
to one. Unlikely, to be sure, but not outrageous.
It should happen about once per season.

The probability that the batter gets twenty or
more hits in his fifty times at bat is 0.085, which
is about the same, so his chances of batting .400
in this stretch are just about the same as batting
.200. Then, of course, he will be said to be on
a streak, which should also happen an average of
once a season. Fluctuations are not unusual in
random events of this sort. One shouldn't draw
too many conclusions from them if they are within
a reasonable range of the expected value. Reasonable range means plus or minus a small multiple
of $\sqrt{N}$, where N is the expected count.

For this particular example, the hitter will be
in his "normal" range a bit more than four times
out of five. If we were dealing with a short season it would be unusual for him to be far out of
range, but the fact that the baseball season now
consists of 162 games means that there will be
many stretches of a dozen games or so, in any
one of which he can deviate from his normal average. Every now and then, for such a stretch, he'll
be declared to either have a "hot bat" or be in
a "slump," whereas, in fact, it is entirely due to
chance.

Of course, one has to be reasonable. If our
batter got no hits at all in his fifty times at bat,

we should be suspicious. The odds against that happening purely by chance are worse than fifty million to one. Something is terribly wrong if it does happen.

This kind of thinking is relevant to all such games. Baseball (America's national game) is a favorite for statistical analysis, and it is in fact found that "batter's slump" is an illusion, but one in which nearly all baseball players, managers, and sports writers believe. In basketball, the number of shots a given player takes during a game is smaller, so the random fluctuations have a larger relative effect. Yet most players, coaches, and fans believe that a player can have a "hot hand," and plays will be devised to get the ball to him, because he is "shooting well." And we've said nothing about winning streaks at the races, or at the roulette or dice tables, or at other such games of chance, where fortunes have been made and lost in pursuing one's lucky streak. To say nothing of the stock market, where nearly all statistical studies show that the major factor in its short-term behavior is random fluctuation.

Now consider a concrete example of the sort that often appears in the press. Two towns we'll call Aardvark and Zyzomys are sociologically very similar, and have the same populations, but Zyzomys is downwind from a giant factory that produces zithers for the mass market. Over a ten-year period it is observed that 100 babies born in Aardvark are tone deaf, while the number for Zyzomys is 110. Should one blame it on the factory and ban the manufacture of zithers, except under rigid government regulation?

Or insist on a major soundproofing of the zither factory, to shield the innocent Zyzomians from the cacophony of zither testing? The answer is no (even ignoring the un-questioned social benefits of the manufacture of zithers), since the square root of either of these numbers is about ten, and one would normally expect random fluctuations of that sort. If Zyzomys had 150 cases of tone-deafness the difference would be much more than ten, would be harder (virtually impossible, the odds being nearly a mil-lion to one) to account for by statistical fluctuations, and would be regarded as "statistically significant." Whether the tone deafness should be blamed on the manufacture of zithers is another matter, since there could be other, previously unnoticed, differences between the towns. This is the excuse used by apologists for the tobacco industry, who deal with the mass of evidence that cigarette smoking causes lung cancer by saying that "statistical association doesn't necessarily imply causation." Though they are technically right, the other evidence for causation in that particular case is overwhelming. Not so for zithers.

Thus, we speak of the "significance" of data, using the square roots of the numbers of events to decide whether the differences we observe are meaningful. Of course the criteria can be made much more precise, but this is enough for now. For the record, biologists have fallen into the habit of regarding data as statistically significant if the odds against the observation having occurred by chance are greater than nineteen to one, for a probability less than 0.05. There is no special rationale for this choice; it is completely arbitrary, but is so ingrained in the biological culture that it is rarely questioned.

This kind of analysis is also important elsewhere. This chapter was first drafted in the midst of a presidential campaign season, and we sometimes saw the results of straw polls announced more or less as follows: "In a survey of 100 registered voters, candidate A has a slight lead over candidate B, with 52 percent of the vote." Since the difference of a few potential voters could easily be a random fluctuation, we know that the so-called slight lead has no significance at all. In fact, the following week may show the same candidate trailing by 2 percent, in which case it might be proclaimed that it is doubtless because he was seen in public with a woman not his wife. Equally insignificant, at least statistically. The more deeply we get into the probabilities of particular calamities, the more troublesome this problem of small numbers will become, especially when there is a large natural incidence of the event in question. Some effects will be just at the square-root-of-N level, and it will be unclear whether they should be treated as real.

Four classes of risk were mentioned earlier, and this is the time to ask how their probabilities are really estimated. It will become increasingly difficult as we work our way through the list.

Familiar high risks

This is the easiest of all, since the bad actors—automobile accidents, aircraft accidents, even murders—show up so often that the statistics are good, and the societal risk is easy to estimate. We did an example in Chapter 2, and need add little to that. The statistics are used directly to determine the probability of an event, and the statistical uncertainties in the estimates will usually be governed by the square-root-of-N rule. The most frequent events will therefore be (relatively) the best known.

Risks of low probability and large consequences

Though major earthquakes were used earlier as prototypes for this kind of risk, they have only a tenuous relation to technology. True, a great deal of scientific effort is going into earthquake prediction and earthquake-resistant design of buildings and other structures, but the source of that risk is not the subject of this book. Nor is dam collapse, threatening though it may be. (Historically, one dam in three hundred fails when it is first filled. Others fail later.)

Structural failure of large objects is, however, a subject one cannot completely overlook, even though such failure is nearly always traceable to human error in design, construction, or maintenance. Every now and then these errors can be considered failures to understand the underlying technology. Here is a case of such a failure.

In 1940 the Tacoma Narrows Bridge, across the narrows at the southern end of Puget Sound, was dedicated with great fanfare as the third-longest suspension bridge in the world. It was over a half mile long, a span exceeded only by the Golden Gate Bridge across the entrance to San Francisco Bay, and the George Washington Bridge, connecting New York and New Jersey over the Hudson River. Each of these had been completed within the previous ten years; that was an active era of suspension-bridge building. Needless to say, this new bridge was carefully designed by a leading designer of bridges, and was a beautiful structure.

Only four months later, in a moderate wind, it collapsed. The films of its collapse, which happened over a period of several hours, are among the most dramatic one can imagine—they show enormous oscillations and twisting of the bridge, flapping of the giant suspension cables as if they were children's jump ropes, and frightened drivers

trying to keep their balance while scrambling to safety,
after having abandoned their cars. The event occurred
over such a long period that everyone on the bridge was
able to escape, and no lives were lost, except that of one
pet dog, who was left in an abandoned car. This kind of
large-scale gyration had never before happened to a suspension bridge, and indeed one's main impression from
the films is of the extraordinary resiliency of the structure.
The bridge took monumental abuse before it finally suc-
cumbed, and was, but for one crucial point, well designed
and constructed.

The inevitable investigation uncovered a number of
things, among them the fact that the unlucky agent who
handled the $6 million insurance policy on the bridge—
about the cost of construction—had pocketed the premiums. Presumably he knew in his heart that bridges don't
collapse and that insurance was therefore a waste of the
taxpayers' money. He ended up in jail. But the technical
investigation, conducted by a committee largely composed
of eminent bridge experts, began by finding no fault with
the original bridge design, and the governor of Washington even entertained for a short time the thought of saving
money by rebuilding from the old blueprints. Fortunately
that thought didn't last long, wisdom prevailed, the bridge
was redesigned and rebuilt in 1950, and it still stands, at
least at this writing. Nothing is guaranteed.

What happened? A vortex phenomenon well known to
designers of airplanes and ships, but apparently not known
to the bridge designers, developed to a disastrous level at
the first high wind. The vortices are called Von Kármán
vortices, after a famous aeronautical engineer, and the account of how he intervened in the bridge episode is told
delightfully in his autobiography. The vortices can be seen

on the surface of the water by anyone rowing a boat or
paddling a canoe—they are the little whirlpools that form
right behind the oar or paddle. The wind caused similar
vortices to form behind the bridge. On that fateful day the
vortices pushed the structure at just its so-called resonant
frequency, the oscillations got larger and larger, and the
bridge finally collapsed. We use resonance when we push a
child on a playground swing—if we time the pushes right,
we can build up quite an amplitude. On that day, the
wind timed it right.

So even experts sometimes make serious mistakes in
principle (not operational errors, which are a different matter), but the probability is very hard to estimate. We do
know from experience that it happens, so we never assume
that people are perfect. To know just how imperfect they
are is one of the more difficult risk-estimation problems.

But this has taken us from the problem of estimating the probabilities for these rare but occasional events.
There are two approaches, empirical and analytic. The
empirical approach is what the term implies—collect all
the data available on the occurrence of comparable events,
and make what is euphemistically called an informed estimate. The analytic approach is, well, more analytic. In
its most developed form it is called probabilistic risk assessment, or PRA, but there is something in that name
that seems to bring out the worst in people. PRA is the
assessment of risk, taking probabilities into account.

The basic idea is that very few disasters appear out
of the blue; most are the culmination of a sequence of
more minor events that don't always end badly. The final
misfortune may be rare, but the events leading to it not
so rare, and we may have useful data on them. If not,
there may at least be experts able to make good estimates.

Therefore, if (and there's the catch) we know most of the ways the final event can occur, we can put together the chances for each of the intermediate events, and make a decent estimate of the final probability. Inevitably there is uncertainty, but it is always better to know something than to know nothing.

As an example of the underlying logic, consider the chance of a perfect game in baseball—a game in which the pitcher retires all twenty-seven batters facing him. That is a very rare event, though a pitcher has about a 70 percent chance of retiring any given batter. Let's estimate the chance of a perfect game—the rare event—from the probability of retiring a single batter, assuming that the batters are independent. That means that the chance of retiring a given batter doesn't depend on what happened to the last one. We have only to use the fact that the probability of retiring one batter is 0.7, of two in a row 0.7×0.7, or 0.49, etc., twenty-seven times. The answer is one chance in about 15,000, or odds of about 15,000 to 1 against a perfect game, for each pitcher.

How good is that estimate? Over the last few decades the major league teams have averaged about three thousand games each year (counting each game twice, for the two pitchers, each of whom has a chance), so we'd expect a perfect game about every five years or so. In fact, over the last forty years, the "modern era," there have been exactly eight perfect games, so our estimate is right on. Such calculations do work.

By knowing the underlying probabilities for the individual contributors (the single put-outs), assuming they are independent of each other, and knowing how they fit into a sequence leading to the rare event (the perfect game), it is possible to estimate the probability of the latter. This was an elementary example because the sequences are simple and completely known—the pitcher has to retire the first batter, then the next, then the next, and on through the ninth inning. There is no other way to the end result. The answer turned out to be surprisingly close to actual experience, and needn't have been—remember the square root rule. We were lucky.

This sort of calculation will work on most long runs of good or bad luck, like winning streaks or losing streaks. There is one authentic exception in sports, Joe DiMaggio's fifty-six-game hitting streak in 1941. No one else has come close. Stephen Jay Gould has argued that, even given DiMaggio's great skill at bat (a lifetime batting average of .325), the streak was an unusual stroke of luck. The odds against *some* good batter doing this *some* year are about a hundred to one. But there is a familiar lesson there too— a low probability is not a zero probability. The event *can* occur.

Take a slightly harder case—say a wing falling off a commercial airplane in flight. This is more complicated, still a rare event, but with more ways to happen. We would begin by listing the various accident paths, knowing in our hearts that we can't think of everything. We would think of the wing being overstressed by a flight into extraordinarily bad weather, and estimate the probability that that might happen. We would think of maintenance errors that overstress the wing spar (it was improper maintenance that cracked the engine mounts on the DC-10 that crashed in Chicago in 1979), and estimate the probability of that sort of error. We would ask for the flight history of the airplane, since the accumulation of metal fatigue leads to embrittlement and the growth of cracks, which can, in turn, lead to structural failure. (All metal components have cracks which develop and grow with age and use, but most cracks don't lead to structural failure until they are large enough and are subject to sufficient stress. Repetitive stress is worst, and led to the failure of the fuselage on an Aloha Airlines flight in 1988. As this is written there seems to be an epidemic of cracked structural members in our aging fleet of commercial jet aircraft.) Were we smarter than people were at the time, we might even think of the flutter that caused the failure of a number of Electra wings when that aircraft was first introduced into service—a phenomenon not unlike the one that did in the Tacoma Narrows Bridge. And so forth.

After we have listed all the paths leading to structural failure, and have assigned probabilities to each path, we can add and multiply the probabilities to estimate the overall probability that the wing will fall off. This will be called a "fault tree," because a diagram picturing all the different paths and their associated probabilities looks like a tree. A side benefit of constructing a fault tree is that we are forced to think clearly about the interactions of the various parts of the system—hand waving is useless. Probabilistic risk assessment involves construction of fault trees and event trees (which are next), to develop an overall estimate of the routes through which an event can occur, and the associated probabilities. An event which is rare because many things have to go wrong to lead to it can still be treated, because the individual contributors to the sequence may not be so rare.

An event tree is the inverse of a fault tree. In the latter we focus on the final occurrence, and list the ways to get to it; in an event tree we start with the event, and list its possible consequences. A typical probabilistic risk assessment begins by constructing event trees. For an airplane, one event tree might begin with an engine failure. Did it happen at low altitude? How good was the pilot? Could the engine be restarted? At each stage, there would be a certain probability that things will get worse or better, and that can be carried to the next scene in the drama. It is like thinking out the consequences of a chess move. Most sequences will end with no one hurt, but there will be a certain probability of disaster, which will by then have been estimated.

Many of these probabilities will depend on the operation of other parts of the system. Are the engine instruments working? Is the electrical system functioning? Is the hydraulic system intact? For each of these there is a probability, and the way to estimate the probability is to construct a fault tree for each item. Thus the fault trees become the means by which probabilities are assigned to the various branches in the event tree, and the composite is called a probabilistic risk assessment.

The procedure is conceptually simple; in practice it can be extremely complicated. At its highest level of current development PRA has been applied to estimating accident risks in nuclear power plants—more in Part II— where thousands of potential accident sequences have been evaluated. The owner of a nuclear power plant who wishes to have a PRA done for his plant had better set aside a few million dollars for the job. Yet it is the best way known to estimate the probability of failure of such complex systems, without actually experiencing the disasters. That is the whole point.

There are intrinsic problems. You can't think of everything; you just do your best. That is called the completeness problem. If there are humans in the system, like pilots or engineers, the art of predicting their behavior under stress is less than perfect. That is called the human factors problem. Sometimes component failures don't occur in a neat order—an earthquake knocks something off a shelf, smashing a control panel and starting a fire while ripping the fire hose in half, at the same time caving in the roof of the fire station. That is called the common cause (or common mode) failure problem.

There are many such difficulties, none of which make the job impossible. They simply increase the uncertainty in the final "bottom-line" answer, but an answer with uncertainty is better than no answer at all. This kind of assessment is now common for nuclear reactors, and is spreading into other areas.

A probabilistic risk assessment can be used—and is the only way—to estimate the probability of very improbable events, things that might happen, say, once in a million years. When a PRA is used that way, and the results are made known, one learns quickly that the vast majority of people, including otherwise quite sophisticated scientists and engineers, have no feeling whatever for what is meant by "one in a million." The Nuclear Regulatory Commission has suggested that a reasonable objective for nuclear safety is a chance of one in a million per year for a major accident in a commercial nuclear power plant, and has had trouble from both sides. Some are complacent because they believe that such low odds mean that it just can't happen, and think the whole thing silly. Others believe that any chance is too much. (Even the Commission seems to be in the first group, mentioning the one-in-a-million number in the same document in which it declares that its policy is to prevent such an event from ever happening.) If members of the second group seek a rationale for worrying about such low probabilities, they point to the fact that those odds are not far from the odds of an individual winning the lottery, and someone does win. The difference of course is that millions of people buy lottery tickets, and there is all the difference in the world between *someone* winning and *you* winning. It is like being told that you are one in a million—exhilarating until you realize that it means that there are hundreds like you in the country.

The interpretation and understanding of such small probabilities lags far behind our ability to calculate them. A probability, however small, means just what it says, and a probability of one in a million means that something will happen *on the average* once in a million tries, if it has been calculated correctly. It doesn't mean it will never happen, nor does it mean that it is certain to happen tomorrow.

Events that have never occurred at all, yet

Here the probabilities are so small that we are reduced to almost pure guesswork. What is the probability that benevolent aliens will land on earth before the next election, and straighten out American politics? Mighty low, but it would be folly to assign a number to it. What is the probability that the next big earthquake on the San Andreas fault will cause California to slide into the Pacific Ocean? Again, mighty low—sorry, folks. What is the probability that the accelerating increase in the population of the earth—it has doubled in the last forty years, and is likely to double again in the next thirty—will lead to a major disaster for the human race within the next fifty years? Very high, but it is social dynamite to even think about solutions, so we don't.

Those are cases in which we can make reasonably informed guesses. They may be little more than that, but are better than nothing. There are others, however, which will yield to a PRA even though the probability is truly infinitesimal. They are like the perfect game.

One far-fetched example might be the probability that a large meteorite falls on a populated city in the United States in the next ten years. There are about forty metropolitan areas in the United States with populations over a million, and with areas ranging from about twenty-five to a few hundred square miles, for a total of very roughly three thousand square miles. The surface area of the United States is about three million square miles, so a meteorite that hits the United States has about a chance in a thousand (if it hasn't been thrown by an extraterrestrial pitcher) of landing in one of our cities. The largest meteorite in recent times produced the crater near Winslow, Arizona, about four thousand feet across and six hundred feet deep. That happened, we think, about fifty thousand years ago, so a reasonable estimate might be that that sort of thing happens every hundred thousand years. We are now ready to make an estimate based on the concatenation of the probabilities; a chance in ten thousand that there will be a big meteorite in the next ten years, and a chance in a thousand that it will land in a big city, for a total probability of one in ten million, or 0.0000001 for the event. It is not impossible, and *could* happen, but don't lose any sleep over it.

More to the point in this category are some items treated more thoroughly in Part II, such as nuclear winter, melting of the Greenland icecap through the greenhouse effect, leakage from nuclear waste repositories, and other technological risks that have a long lead time and low probability. Also in this category, strictly speaking, is a major nuclear accident in the United States—it hasn't happened yet, at this writing, but the probability is reasonably calculable, and it is inevitable, given time. For all of these, the methods just described are applicable, and there is really no difference in method between rare events that have happened and those that haven't yet.

Risks buried in a background of "natural" occurrence

We've listed some examples earlier—chemical carcinogenesis, radon, pesticides, contaminants in drinking water, and so on. The problem is one of distinguishing small additional effects, to see if they are real and, if so, how important. (At this writing we are in the midst of a national mini-panic about radon. Even supermarkets are selling so-called radon detection kits.) The main tool for analysis here is the square-root-of-N rule, described earlier in the chapter, and elaborated in Part III. It sets a limit on our ability to measure small changes in a natural background.

There are, however, other tools. For example, the type of cancer caused by long inhalation of asbestos fibers is microscopically different from that resulting from other insults to the lungs, like smoking (though there is evidence that these are synergistic), so it can be detected in a background. If the damage bears the specific trademark of the cause, it can be identified even if it's a small increment. (Think of a ballistics test on a murder weapon.)

Similarly, some risks lend themselves to controlled experiments, normally on laboratory animals. Animal lovers were revolted some years ago by experiments in which beagles were used to assess the effects of smoking on people, but a great deal was learned thereby. The same is true of animal experiments involving other potentially dangerous materials, because they can be conducted in a situation in which the effects of the material in question can be isolated. Indeed, one of the real handicaps to AIDS research at present is the shortage of animal species susceptible to the virus. AIDS seems destined to be one of the greatest killers in the history of the human race, but the risk is not technological, and at least at this moment (late 1989) people have still not come to grips with its impending magnitude.

For this category of risk, then, the major tools for assessment are careful statistical and epidemiological analysis, combined with laboratory and theoretical studies of the phenomena. Where the latter are absent, we are reduced to the former, and risk assessment is a much more difficult job. There will be examples in Chapter 12.

6

The Management of Risk

There are two basic strategies for the management of risk: prevention and mitigation. The first means to lower the probability of the unwanted event; the second means to render the consequences less unpleasant when it does happen. (Storm cellars for tornadoes are purely mitigative; no one can yet steer tornadoes. Cloud seeding, in the hope of changing the meteorological conditions that lead to tornadoes, would be preventive if it worked.) In the response to disease, prevention is called prophylaxis and mitigation is called therapeutics. We'll deal with mitigation through an example in Chapter 15, the emergency management of a nuclear accident, and devote this chapter to prevention. (Strictly speaking, there is no such thing as prevention; we mean only to lower the probability.)

Much of the preventive effort for technological risk is channeled into regulation—we Americans seem to have a mystical faith in the ability of regulatory agencies to find a way to make our lives safer without seriously degrading them. How often do we complain when a harmful drug (like thalidomide) slips through the regulatory maze, and how often when the seemingly eternal testing process keeps touted cures off the market? Even laetrile, almost universally regarded by experts as an unconscionable fraud on the desperately ill, has its share of vocal supporters. The scene is being replayed for purported AIDS cures, some of which may conceivably carry some promise. But which? And eager customers abound—suffering people will clutch at straws. That choice between denying us good and protecting us from bad or worthless is the principal dilemma of regulation.

Some preventive effort isn't regulatory, but also not typically directed against technological risk. Immunization has largely wiped out some formerly widespread diseases, and public health measures even more. We now believe that smallpox, once a dreaded disease (over 200,000 cases in the United States in the epidemic of 1920–21), has been eliminated from the earth. That was vaccination. The plague, which killed an estimated quarter of the population of Europe in the fourteenth century, is a classic example of a public health success. It was spread to people by fleas from infected rats (though person-to-person transmission does occur), and it was necessary only to break the chain of infection somewhere to reduce this threat to minor status. Similarly with malaria and mosquitoes, but only in some areas—malaria is still a major problem in many parts of the world.

But regulation is our preferred weapon against technological risk. Some regulatory agencies have existed for a long time, but we've gone all out creating them in the past couple of decades. Here's a list: the Occupational Safety and Health Administration of the Department of Labor (OSHA), the Consumer Product Safety Commission

(CPSC), the Environmental Protection Agency (EPA) and its Offices devoted to water, air, wastes, pesticides, radiation, etc., the Food and Drug Administration (FDA) of the Department of Health and Human Services, the Nuclear Regulatory Commission (NRC), the Federal Emergency Management Agency (FEMA), the Bureau of Alcohol, Tobacco, and Firearms of the Treasury Department (ATF), the Mine Safety and Health Administration of the Department of Labor (MSHA), the National Highway Transportation Safety Administration (NHTSA) of the Department of Transportation, the Federal Aviation Administration (FAA) of the Department of Transportation, and its watchdog, the National Transportation Safety Board (NTSB), which is an independent agency, the Drug Enforcement Administration (DEA) of the Department of Justice, the Coast Guard (USCG), now part of the Department of Transportation, the brand-new Defense Nuclear Facilities Safety Board (DNFSB), the legions of local and state agencies, and we've hardly begun. Each of these has a statutory foundation which establishes its realm of responsibility and authority, at the same time giving it teeth. (Some have only grinding molars, others fangs.) Most have a network of advisory committees, many statutory, whose function is to keep an eye on (and of course help) the regulators. A regulatory agency usually has a host of loosely related laws to enforce (once created it becomes a natural repository for new responsibilities), and an even longer list of internally generated rules. Sometimes the laws are mutually contradictory, and the domains of regulatory agencies overlap. The boundaries of authority for different agencies are the scene for turf fights, either in the courts or in the Congress. The different agencies are under the jurisdiction of different committees and subcommittees of the Congress, which are therefore drawn into the territorial warfare. Of the organizations listed, about three-quarters were created in the 1970s, though a few can trace their ancestry much further back. They are all twentieth-century creations.

Each agency has a birth saga, usually an incident that attracted public attention and demands for protection. Protection of the citizenry is, after all, a principal function of government. (William the Conqueror, knowing his priorities, prohibited the adulteration of beer.) To pick just one example from our list, the FDA, despite a genealogy tracing back in one way or another to the Food and Drug Act of 1906, got its teeth with the passage of the Food, Drug, and Cosmetic Act in 1938. (In hindsight this was a curious combination of subjects for one agency, though, as David Bodanis points out in his delightful book *The Secret House,* when the New York Board of Health considered banning lipstick in 1924, it was not for the benefit of women but for fear that the lipstick might poison the men who kissed the women who wore it. Some activities are risky for unexpected reasons.) The new act put the burden on manufacturers to prove the safety of new drugs. The catalytic event was the fatal poisoning in 1937 of about a hundred people, mostly children, who had been given a syrupy version of sulfanilamide, the first of the broad-spectrum wonder drugs. Alas, the syrup was formulated (out of ignorance, not malice) from a base of diethylene glycol. As with all poisons, there is a damage threshold, and taste testing by an adult chemist (who later committed suicide) was insufficient to protect children given therapeutic doses. So a law that had been waiting in the wings took center stage, and was quickly passed. There is a similar story in other cases.

Getting back to the list of agencies, this *embarras de richesses* creates business for the establishments and institutions that feed data and analyses to the regulatory agencies. These include the National Academies of Sciences and Engineering (NAS/NAE), the Centers for Disease Control (CDC), the National Institute for Occupational Safety and Health (NIOSH), the National Institute for Drug Abuse (NIDA), the Office of Science and Technology Policy (OSTP) in the Executive Office of the President, the Office of Technology Assessment (OTA) in the Congress, scores of advisory committees, universities, nonprofit research organizations, profitable research organizations, consultants, contractors, writers of books, witnesses in hearings, concerned citizens, unconcerned citizens who enjoy the publicity that goes with opposing something, lobbyists, and so on. Support for regulatory agencies is not the only work of these organizations, but it does supply part of their gainful employment. And even that should not be interpreted in a negative sense, since this subculture provides much of the internal consistency of risk regulation.

This last point requires some expansion, for more reasons than risk. If each of the regulatory agencies functioned on its own, setting standards, establishing criteria, and making rules, we would have a government consisting of so many uncoordinated parts that no one would ever know what was permissible and what was not. There has to be communication among the agencies that protect us, just to provide consistency and rationality in the face of conflicting and uncertain criteria. There is even a "state of the art" of risk assessment—that is what this book is about—that ought to be applied even-handedly.

Here the risk subculture plays a role. There are conferences, speeches, papers, reports, and yes even parties, at which information is exchanged, after which the practitioners, like modern-day Johnny Appleseeds, propagate a kind of consistency throughout the community. All sciences work like this, formally or informally setting consensus standards for the practice of the art. Anyone who has ever tried to learn what is known about some esoteric topic—say the rheology of sewage slurries or the syntax of Linear B—has been hard-pressed to find anyone who knows anything. Having found one such person, however, he will have found them all. (That's the meaning of the word "esoteric.") Knowledge lives through the activities of interacting communities of informed and inquisitive people, who occasionally leave their droppings in written form.

It is easy to overlook the fact that this cross-fertilization plays an essential role in the coherence of government. In 1984 the White House Office of Presidential Personnel issued a proclamation that it would no longer be permitted for a single individual to serve on more than one Presidential Board or Commission. If obeyed, this would have summarily isolated each function from the next. The reason: "Because of the large number of qualified candidates ... this will provide a greater number of opportunities for those people anxious to serve the President." In other words, spread the jobs among the faithful. The edict was generally ignored. (There is a widespread misconception among people who have had little experience with the workings of government that issuance of an order means that what was ordained will actually come to pass. There are in fact so many laws, rules, and directives that the vast majority have no impact whatever. Active enforcement or active lawyers make a rule, not the rule itself.)

Another glue in setting our regulatory agenda is the

press, including television, whose interpretation of risk can be decisive. The United States is, for better or worse, a democracy, and we judge importance through noise. This author does not share the common view that reporters spend night and day in fiendish schemes to distort words and misrepresent facts, yet much of what we read in the press, especially with regard to risk, is less than accurate. Nearly all reporters would prefer, other things being equal, to do a good and accurate job than a bad one, but other things are rarely equal. In technical matters the facts may be in dispute, they may not be clear, the subject may be recondite, people may be covering up, the investigation may be incomplete at press time, etc., and the reporter has to do his job. He is, under the best of circumstances, a journalism graduate, and only the biggest and finest (not the same) newspapers can afford a science writer.

Since our news media are, with rare exceptions, commercial, a good story is more likely to preempt the expensive facilities than a bad story. (Whether news or soap opera, advertising pays the freight, and it goes to the programs that people watch, and the newspapers and magazines people buy.) We therefore emphasize flashy events and rare diseases, to the point that people come to think that what is being reported fairly represents what is happening. There is no malice or misbehavior here, only a normal tendency for people to be interested in that which is interesting, and a news medium that acknowledges that fact is more likely to be successful than one that does not.

The reason for mentioning all this in a chapter on regulation is that the people in our regulatory agencies (to say nothing of the Congress) also read the press and watch television, and are guided in their priorities by what they see and hear. We respond to the risks we perceive—what else can we do?—and are especially concerned if there is uncertainty. Since we are a democracy, our government cannot behave very differently from any commercial organization catering to our whims. A congressman or city councilman or county supervisor who tells his constituents to go packing because they don't know what they are talking about will not last long in office.

Finally, all too often, there sit above all these regulatory agencies and other forces the courts. The courts are the proper arbiters of disputes about procedures, authority, and the interpretation of law and the Constitution, but they also tend to have the last word in the settlement of technical disputes. The concept boggles the mind, since there is nothing in the education or experience of a judge (or indeed a jury—this author's personal experience is that people with technical education are automatically dismissed during jury selection) that prepares him or her for that role. Often there isn't even a well-defined or generally accepted answer to the technical question at issue. Scientists are not troubled by the idea that a question may simply not have an answer, but judges and lawyers are. They tend to think that the witness has either not tried hard enough, or that someone is covering something up. (In one case, a U.S. judge wrote, "Our present system of review assumes judges will acquire whatever technical knowledge is necessary as background for decision of the legal questions . . . ." As pipe dreams go, that is first class.)

In another case a judge discounted any testimony that had to do with the probability of an event, saying that to do otherwise would mean that he was gambling, and "this Court is not a bookie." He was overturned by the Supreme Court.

With some ingenuity and dedication we could probably find a worse way to settle technical disagreements. We do it this way because the Congress, which of course creates the laws, is rarely clear on how much safety is sought, and therefore writes the laws ambiguously. No one is villainous or inept—these are difficult questions, with no political consensus on our goals and objectives. On the rare occasions when the Congress is specific, as we will see it was with carcinogenic additives to foods, the effort backfires.

It would be hopeless (and dull beyond reason) to try to describe in detail the responsibility and *modus operand* of each of these agencies. Also, sweeping generalizations would do little good. Those matters are best dealt with by example, and that is the function of Part II, where some real cases will be traced through the regulatory system.

There will be one focal point. The almost universal obstacle to rational regulation is the failure of the laws to specify an acceptable level of risk, and what we are willing to pay to get there. This leaves the regulatory agencies to do more than just regulate—they have to exercise judgment in setting the standards against which they regulate. They have been made into both enforcers and rulemakers. This means that the doctrine we learn in our high-school civics classes—that those who make the rules are best kept separate from those who enforce the rules—breaks down in the case of regulatory agencies.

A typical agency thus splits its job in half. First it must judge the intent of the law under which it functions, and then formalize that judgment as a set of rules. Then the rules can be enforced, rather than the law. This arrangement is also better for those who are regulated. A general principle of law requires that someone held to a requirement ought to know what the requirement is. Life

is not supposed to be like a Kafka novel, in which one can
be accused of a crime but not told which one. Unfortunately, the arbitrary standard can develop a life of its own,
and become the purpose itself, rather than the imperfect
expression of the purpose.

Thus, excessive speed on the highway is dangerous,
but what is excessive? The laws in most states say something like "too fast for conditions," but that just sidesteps
the question. Enforcement makes it necessary to define
a particular speed as the *prima facie* limit, meaning that
speeds in excess of that are presumptively or evidently
dangerous. Then the speed limit becomes an enforceable
surrogate for safety.

By the same token, we have, in most states, a blood
alcohol limit of 0.1 percent alcohol as presumptive evidence of drunkenness while driving. Typically a police officer has some discretion between 0.05 percent and 0.1 percent. Below 0.05 percent the driver is presumably sober.
Fortunately, we have slightly stricter federal standards for
pilots—less than 0.04 percent blood alcohol *and* at least
eight hours since drinking. It is therefore possible to be
a sober driver and a drunken pilot, at the same time. In-
toxication on drugs other than alcohol is treated much
less precisely, for pilots, drivers, and other critical workers, and we haven't yet agreed on a comparable standard.
That makes enforcement much more difficult. (The Federal Air Regulations for pilots are precise about alcohol,
but for other drugs they simply state that no one should fly
an airplane when affected enough to impair safety. Who
knows exactly what that means?)

The 0.1 percent alcohol limit is arbitrary, serving only
to make law enforcement easier by relieving the authorities
of the need to prove that the intoxicated driver was *really*

impaired. In Washington, D.C., the nation's capital, the limit used to be 0.21 percent, but is now 0.1 percent. At this writing four states (California, Maine, Oregon, and Utah) have a somewhat stricter standard of 0.08 percent blood alcohol, and Georgia is more tolerant at 0.12 percent. The American Medical Association is encouraging the adoption of a 0.05 percent standard in all states. This latter illustrates one of the general features of arbitrary regulatory standards—political considerations preclude relaxing them, and there is constant pressure to make them more restrictive. This process is known to cynics as ratcheting; the standard was arbitrary in the first place, and is neither more nor less arbitrary when tightened.

In the face of this arbitrariness the pragmatic definition of sufficient safety becomes a level at which the political pressure dies down or the budget runs out, which is almost the same thing. Since society's resources are limited, attention lavished on risks that either do not exist or are trivial inevitably reduces efforts to deal with real risks. Or, for that matter, to feed the hungry.

New administrations sometimes set out to "put the house in order," since nearly everyone recognizes that there is more self-perpetuating bureaucracy than appears necessary for our well-being. The Reagan administration was no exception when it took office in 1981, and the President signed an Executive Order requiring that any executive agency that issued a major regulation certify that the benefits exceed the costs. How could anyone object? The words were unexceptionable: "Agencies shall set regulatory priorities with the aim of maximizing the net benefits to society ...." In other words, certify that you are doing more good than harm. He went even further down the yellow brick road: "Administrative decisions shall be

based on adequate information concerning the need ...,"
and "Regulatory action shall not be undertaken unless the
potential benefits to society outweigh the potential costs
to society," and again, "Regulatory objectives shall be chosen to maximize the net benefits to society." And on and
on. The order also stipulated that the required Regulatory
Impact Analysis take into account "any beneficial effects
that cannot be quantified in monetary terms," presumably
to deal with questions like beautiful sunsets and redwood
trees.

The predictable result was that the administration was
accused of too cold-blooded and heartless an approach to
human health and safety, and the agencies (which are in
the business of regulation) found many ways to get around
the order. It is still mentioned, has not been withdrawn,
and is largely irrelevant. A fundamental problem in imple-
menting any cost-benefit strategy, of course, is that those
who receive the benefits rarely pay the costs. Therefore,
though the overall good to society may ideally be maxi-
mized by matching benefits against costs, politics may not
balance the losers against the winners. Some opponents
of new electric generating capacity take the position that
the only beneficiaries are the electric utilities, who make a
profit from the sale of electricity, and that the electricity
does the consumer no good at all. And regulators are in
the business of regulation, often regardless of the cost to
society.

This is the cornucopia theory of regulation—if it is
public money that is being spent, it is so hard to identify
the source that there is very little incentive to economize.
The beneficiaries are visible, but not the payers. Anyone
who asserts that a public good ought to be judged not only
in terms of whether it is good, but also in terms of whether

it is *sufficiently* good, gets into trouble.

In the mathematical theory of games there is a concept called the "zero-sum game." A zero-sum game is one that a player can only win at the expense of another player—the players can't all win. Poker is such a game, and everyone who plays recognizes the fact that he is playing *against* the other players. Very few of us recognize that expenditure of public funds for the public good, including regulation, has many features of such a game. The resources come from somewhere, from somebody, and deny us the option of doing something else, though it may be impossible to identify just what.

The issue of determining a proper level for regulation, given the costs and benefits, is the most perplexing problem. A license to operate a nuclear power plant is issued after a certification that the plant can be operated "without undue risk to the health and safety of the public." That is the standard. No one has ever defined how much risk is due the public, and the question, when raised, is normally met by an embarrassed silence. This theme pervades all regulation.

7

The Value of Life

How can there be a subject called the value of life? Life is well known to be priceless, and that ends the discussion. Therein lies the problem: how hard should we work at limiting risk to a reasonable level, if we can't decide what is meant by the word "reasonable"? As a society, we really can't afford to act as if life were truly priceless.

To put the point in perspective, we can try to estimate what a life would be worth in crass economic terms, ignoring all social, moral, and spiritual matters. Use the United States as an example. In 1986 our gross national product was about $4.2 trillion (million million), for a population of about 240 million, so about $17,500 per capita. With a life expectancy of seventy-five years, this suggests a gross product of somewhat over a million dollars per lifetime for the roughest kind of measure of what we produce in our time on earth. Such a measure should not be taken very seriously, for many reasons. Is a person called productive if he strikes oil? If he plants corn, cuts down forests, plays football, or is a rock star? All of these contribute to the gross national product. What about a burglar or a drug merchant? It isn't easy to define productivity.

Suppose all these jobs are called productive, and we again eschew all sentiment and think of a person as a productive machine. As such, he requires a certain level of support services: fuel (food), protection from the elements (housing), maintenance and repair (medical and dental), and so forth, which will eat into his or her apparent productivity, just as they would for a tractor or lawn mower. People require upkeep, and even require entertainment, which machines do not.

A little reflection reveals that the average person more or less breaks even, consuming about as much as he produces in his lifetime. For a growing society there can be a net production edge, but the converse can also be true, and in any case the margin tends to be small. In the United States, as this is written, we are consuming more than we produce, which is why our country is building a larger and larger debt to the outside world. It can't go on indefinitely. As we mentioned in Chapter 3, individuals also vary in productivity as they march through life; people in the middle years produce more than they consume, and the very old and very young consume more than they produce. That is the social contract, sustained by the veneer of civilization, and has as its pragmatic basis (as far as children are concerned) the need for group survival. Many societies, and indeed species, have not survived, a fact that is all too easy to overlook. As mentioned earlier, species survival is the single most important obligation we have toward our successors, and it is not guaranteed.

If we seek a reasonable estimate for the value of life, to use in dealing with risk questions, we won't find it through this kind of cold analysis. Calculations of the sort described above are often used in the courts in dealing with wrongful death lawsuits, but those normally deal with only one side of the equation—a person's potential earning power. The upkeep cost is never computed, which is equivalent to balancing a company's books by discarding all records of expenditures. The value we set on life has little to do with productivity.

Another way to learn the value of an item is to let market forces have their way, using the laws of supply and demand. That wouldn't be much use here, since the principal problem facing mankind is doubtless overpopulation, and oversupply should drive the value down in a free market. We can hardly let that logic govern our approach to understanding the value of life. There is a paradox here, but so be it. We had also better not rely on the cost of production as a measure of value, since it is notorious that people can be mass-produced, cheaply, using only unskilled labor. Neither of these marketplace approaches is much help.

We can, however, ask whether our society's feeling about the value of life isn't realistically expressed through everyday decisions about how much we are willing to spend to limit risk. Decisions on these matters reflect our collective and individual *attitudes* toward the value of life; in a free economy attitudes determine the value of all commodities. The list of highest paid people in the United States in 1987 was dominated by entertainers, with incomes ranging up to $50 million (for a prizefighter), closely followed by a rock singer, so that is by definition the value of their services. Some may feel that this is an absurd distortion of the concept of value, but then few of us don't feel underpaid.

There are clues in everyday behavior to the value we

set on life, of which three are most often mentioned. We pay people a premium to take on risky jobs, our democratic governments have limited budgets for the reduction of life-threatening risks, and our regulatory agencies have developed *de facto* standards that seem acceptable. Obviously, these aren't independent, but they are all measures of the free market value of life when real money is at stake. We are still speaking of the United States—different views are held in different countries. (After the Bhopal disaster, the Indian claimants wanted to try the damage suits in American courts, where life is apt to be set at a higher price.)

Salary incentives are common for people in unusually risky jobs, both as a form of compensation for the risk and as an inducement to sign on. (All salaries, as well as bribes and tips, serve the dual purpose of compensation and incentive.) In principle, it should be possible to find out how people value their own lives by asking how big a salary premium is required to induce them to move into risky employment. Sometimes, of course, the interpretation can pose a problem, as in the cases of combat pay and flight pay for the armed forces, where the increased pay bears little relation to the increased risk, and the environment is hardly that of a free market.

Nonetheless, such studies tend to bracket the estimates of the value of life at between a few hundred thousand dollars and a few million dollars, as seen both by those who take the jobs and those who offer them.

The most serious problem with this approach to estimating the value of life is that there are few jobs that pose a serious threat of loss of life. Some people do earn their living through the *perception* that they are risking their lives (as in auto racing and motorcycle stunting), and indeed their jobs are too risky for some tastes, but most such daredevils die, in the end, of the same afflictions that bedevil us all. Meanwhile they are well paid for their entertainment value, which is enhanced by the perception of risk.

For less risky jobs there are smaller incentives, so they don't yield very good estimates of the implied value of life. Such estimates as there are yield values between a few hundred thousand and a few million dollars, but people's perceptions of small risks aren't very reliable, so the estimates are not very helpful.

The author flies over a hundred thousand miles per year, on commercial airlines for which the passenger fatality rate is about one per billion passenger-miles, so the probability of dying in a work-related airplane accident is about one in ten thousand per year. That is about half the combined risk of being murdered or committing suicide, and commands no salary premium. Even for the more publicized occupational risks, like working in coal or uranium mines all one's life, or teaching (professors occasionally get shot by frustrated students), the actual risk is small.

What about the second approach to setting a value on life—the expenditures governments are willing to make to protect us? There is plenty of evidence here, ranging from expenditures on roads and hospitals to efforts to keep high-level nuclear waste out of the biosphere. Professor Bernard Cohen, of the University of Pittsburgh, has studied a variety of government expenditures, and found that we are inconsistent; in some areas of risk we spend little, in others the sky is the limit.

Highway improvements like guard rails, warning signs, better maintenance, etc., could save lives at relatively low cost—$20,000 to $50,000 per life. These are bargains if a life is worth a million dollars. At the other end, we throw money at some problems without saving very many lives. Cohen's estimate of expenditures on coal mine safety is $22 million per life saved (in 1975 dollars). Of course we should try to save coal miners' lives, but could save many more drivers' lives at the same cost. Miners seem to be more valuable than automobile occupants. It may be that we think of miners as helpless victims, and believe that drivers ought to be responsible for their own safety. Or perhaps there is no drivers' union watching after their welfare. Whatever the reason, the contrast is interesting. There are even more extreme examples.

Cohen also found that the funding of expanded immunization programs in Indonesia could save lives at a cost (again in 1975 dollars) of about $100 per life. When this author first heard this, about ten years ago, he mentioned it to a class he was teaching, and suggested that the class contribute $100, at a dollar a student, to a fund which would be dispatched to Indonesia, to be used to save a life. There were no contributions.

None of this is simple; there are deep ethical questions about the preservation of life per se, if the consequence is more deaths from worse problems in the immediate future. What of famine relief, a moral imperative, if its inevitable consequence is to later doom even more people to death from starvation? One cannot in good conscience hesitate to help hungry people, but in many cases, and in the long run, compassion may do more harm than good.

Another case is closer to home. The Department of Transportation has estimated that mandatory seat belt usage would save lives at a cost of about $500 per life. The cost is low, consisting mainly of the cost of enforcement—

the cars already have the belts—and the lives saved, even though exaggerated by some public advocates, and quite possibly by DOT, are still likely to be substantial. In the abstract, any of us would doubtless vote to save lives at such a low cost, yet mandatory seat-belt laws are not popular. More about this in Chapter 13.

We haven't gotten very far by trying to understand the value of life through the resources that (democratic) governments will devote to its preservation. Values vary from the low tens of thousands of dollars, or even lower, up to billions of dollars (high-level radioactive waste disposal) for each life saved. There are bargains available, but in rather unspectacular forms—highway repair, medical screening, seat belts, etc.—with such options as mine safety at the other extreme. As always, decisions are tied as much to the perception of risk as to the substance. Think of what we would gladly spend to save an astronaut trapped in space, or what we have spent to ransom a hostage held by terrorists. In the latter case our compassion has led us (and we are not alone) to contribute to the support of future acts of terrorism.

A third way to test the value of life is to look at the behavior of those regulatory agencies charged with our protection. These agencies have resources that reflect, however indirectly, the concerns of the taxpayers. Agencies respond to the pressures that are placed on them by the electorate and the body politic, and their behavior ought to reflect our own views about the value of life.

Here we find more consistency than we have a right to expect, especially since the agencies have to formulate their own criteria with little guidance. Yet agencies copy from each other, and learn what the courts, the Congress, and the public will accept. Acceptance sometimes depends on the fact that there is no technology available to bring risk to an "acceptable" level, at any price. Then risk reduction can only be achieved by shutting down entire industries, and we tend to become less idealistic when that is the issue. (We always operate on a high moral plane if the price is right, but lose our zeal when our own resources are at stake.)

A group from Oak Ridge and from Cambridge has analyzed the actual behavior of a sample of regulatory agencies responsible for protecting us from chemical carcinogens, when the decision on whether or not to regulate involved an estimate of the value of human life. There were dozens of chemicals involved, and the pattern was clear. If lives could be saved at an estimated cost of less than $2 million per life, the exposure to the relevant carcinogen was regulated; if the cost was higher it was not. In addition, some slightly risky substances were not regulated simply because the risk was deemed negligible per se. The dividing lines were not clear-cut, but a variety of regulatory agencies behaved similarly in setting out to do their jobs. One shouldn't be too surprised, since the risk subculture mentioned in the last chapter provides a bridge between agencies.

What can we make of all these efforts to value life?

First, there is increasing recognition that one can't simply adopt the "moral" position that life is priceless, and that there is no limit to the resources that must be devoted to its preservation. In a world in which resources are limited (this very world), resources wasted where they do little good will inevitably rob other worthwhile activities, with a net loss of life. People who try to do good everywhere are a handicap to those who try to do good somewhere.

Second, there is a drift away from the idea that the value of life ought to be based on the lifetime earning power of the individual. Though many legal disputes are resolved this way—the family of a successful executive will usually get a larger settlement in a wrongful death suit than will the family of a laborer—government agencies must treat everyone alike. That makes it particularly difficult to find a logical basis for setting a value on life, because the value has to be set in ignorance of anything about the individual.

This last point emerged after an accident at a construction site in West Virginia in 1978, in which some inadequately cured concrete gave way, and fifty-one construction workers died. Over the succeeding years OSHA, struggling to formulate safety standards for such concrete and masonry construction activities, inevitably came face to face with the problem of the value of the workers' lives. The costs of tightened standards were real enough, and would be paid by consumers through the increased cost of construction, but the benefits were in the value of workers' lives saved. Were these lives worth less than those of company presidents, or pilots? In the end, OSHA did the analysis for a range of life values, from $2 million to $5 million, considerably higher than the lifetime earning power of the workers involved, the "human capital" value.

Third, there is little or no consistency at the state and local levels, where life-saving activities are valued in a range so broad that no underlying rationale is apparent. The idea that one can or should do analysis in these matters has just not sunk in very deeply. Local and state governments are exquisitely responsive to the electorate; traffic lights are installed where the citizenry demands them, not where they will save lives.

Fourth, federal regulatory agencies seem to have converged on a *de facto* estimate of the value of life centered near a million dollars, ranging above and below by about a factor of five. There are extreme cases of wild overestimates, but they are few. Of course, when the practicability of a government action depends on whether a life is valued at $999,999 or $1,000,001, the decision is made on other grounds, as it should be. Decisions must sometimes be made which are hard to justify on the basis of logical analysis alone—who uses a calculator to decide whether to go on a date? It is comforting to remember that if a decision is marginal, the arguments one way or the other being about equally persuasive, it probably doesn't matter much which road is taken.

Finally, anyone who tries to set a monetary value on human life is certain to unleash a flood of angry vilification from the self-appointed custodians of everyone else's morality.

8

How Safe Is Safe Enough?

Ay, there's the rub. That is the central question around which all risk management revolves, and it brings together the questions of risk assessment and risk management, two activities that are otherwise best kept separate in our minds. Each is exactly what the name implies, and so far Chapter 5 has been devoted to the former and Chapter 6 to the latter. Management is what is done to limit risk, and assessment is what is done to determine whether the results are satisfactory. They tend to get mixed up.

Risk management can be practiced without risk assessment, just as baseball can be played without keeping track of the batting averages of the players, the earned-run averages of the pitchers, the idiosyncrasies of each, or even of the score. But it isn't possible to do it effectively in the long run. By the same token there are baseball buffs (usually, but not always, juvenile) who keep track of the statistics but have no real use for them. They are amateurs, in the best sense of the word. The risk assessment business teems with people who do assessments with no thought of how or whether they can be used to improve risk management.

This is not trivial. There is meaning and virtue to the collection and interpretation of knowledge for its own sake—that is what pure science and mathematics are all about, and we owe a lot to them—but there is added value when they are used to better our performance in life. It is also possible to do creative and useful things in life with minimal understanding of the underlying body of knowledge—there are successful tinkerers—but understanding helps. There are zealous amateurs in most fields, even risk.

Conceptually, the simplest way to deal with the question of "how safe is safe enough" (HSISE) is to reject the whole idea, and to assert that in every activity we should expend all available resources in the reduction of risk, seeking a no-risk world in the same way that medieval knights sought the holy grail. The chance of success would be somewhat smaller; there *may* have been a grail. Many of the organizations mentioned in Chapter 4 make a public display of seeking to eliminate risk, though they usually deny it in conversation with experts. They recognize that the position is indefensible, but it is at least one that has no internal contradictions; when faced with a risk, one works to eliminate it, and stops only when the resources run out.

Unfortunately, many of our laws have been interpreted by the courts to mean that regulatory agencies are forbidden to balance the social and economic costs of risk reduction in deciding HSISE. The agencies are therefore left with teams of lawyers struggling to defend reasonable compromises, fighting other teams of lawyers arguing for strict compliance with carelessly written statutes.

Some agencies—for example, FDA in cases that don't involve carcinogenic additives to foods—have managed to
get along reasonably well, using plain vanilla arbitrariness
as a means of deciding HSISE. That works, *provided*, and
FDA is simply lucky that way, the enabling legislation permits it. It is painful for a scientist to say this, but we do
sometimes get so involved in arguing technical nuances
that only a less well-informed person can make any decisions at all. Arbitrariness isn't all bad, especially if it
reflects an underlying philosophy.

It is possible to get along without any strategy at all,
by simply bumbling through, making each decision in response to the immediate pressures. This has been called
the marketplace approach, since there is no consistent rationale beyond responsiveness to force. The basic assumption is that people in a free market will pay for whatever
safety they really want, and the best way to find that level
is to figuratively put the whole matter up for bids. It is
comparatively easy to function that way. The FDA sets
its arbitrary tolerances for potentially toxic food additives
by first assembling the relevant data. Except when con-
strained by the Delaney clause in the case of carcinogens, it
takes into account animal experiments, human experience,
the number of people likely to be exposed, and the impact
on the food supply (but not the economic impact on the
producers of that food). Tolerance levels are then set according to its best judgment. There is no exact criterion,
relevant to all cases, and no claim that there is.

FDA is then bombarded with comment and pressure
from both those who think the tolerance is too high—
often consumer groups—those who think it is too low—
often producer groups—and the Congress, and the agency
is constantly embroiled in litigation. Such an openly ar-

bitrary approach has the usual merits and demerits of the free market system, including the demerits of wasted human resources and continued conflict. It is not an unreasonable compromise. The criteria for setting tolerances are kept fuzzy (deliberately, one suspects), but the debate is kept open. Democracy is not supposed to be an effective form of government, only superior to the visible alternatives. At this writing, it is unclear that it is even a viable long-term form of government—a few hundred years of history hardly constitute adequate evidence—but that is a more important subject than technological risk.

A related approach to the HSISE question is to set entirely arbitrary risk standards in advance—without analysis, but according to what the traffic will bear—and then forget that they were arbitrary. For example, EPA recently relaxed the standards on a number of environmental contaminants, including dioxins and lead, in an effort to bring the population risk to a uniform level of one in a million chance of untimely death per lifetime (seventy years). (This arbitrary choice of risk level is being increasingly used throughout government, presumably because it is a round number and is, for most people, synonymous with no risk at all.) EPA was again subjected to abuse from environmental organizations, who are relatively consistent in demanding the most stringent possible regulation of contaminants, and who generally (and in this author's obvious view, unthinkingly) oppose any effort to set quantitative standards. Arbitrariness does invite legal attack.

Unfortunately the arbitrary-standard approach to risk management doesn't take adequate account of the uncertainty in the risk estimates. When we discussed the square-root-of-N rule in Chapter 5, it was to establish that the kind of risk under discussion here—the risk of contracting a low-probability cancer from a particular contaminant, in a sea of cancers which cause 22 percent of our total mortality—is extremely hard to quantify. There is considerable uncertainty in the estimates, which the political players can then use to their own advantage. Those who believe cancer is an unnecessary evil that needs to be stamped out by regulation need only be guided by worst-case assessments of the probabilities, at the limits of the uncertainty bounds.

In the case of nuclear power the Nuclear Regulatory Commission (NRC) has confronted this problem by promulgating a set of formal Safety Goals, which specify precisely how much risk is to be inflicted on the public by the nuclear enterprise—compared to the other risks in life and to the risks of other electricity-producing technologies. Though these goals are entirely arbitrary, they are directly based on a comparison of nuclear risk with other risk. The NRC lacks the statutory authority to compare risks with benefits. So the uncertainty in the risk of nuclear power is compounded with the uncertainty in the comparison risks, and the situation lends itself far too easily to worst-case treatment.

And the uncertainty matters. Every driver knows that if the posted speed limit is 55 mph, it is possible to get away with 60 or 65, if only because there is a bit of uncertainty in the measurement of a car's speed. The law enforcement agencies allow for it, citing only those drivers who are clearly and demonstrably speeding. But what if the uncertainty were not a matter of (say) 10 mph, but were 50 mph? Then the range of uncertainty would be from 5 mph to 105 mph. The speed limit would become pretty useless as an enforcement tool. The nuclear uncertainties are even bigger.

In addition, there is a certain illogic in comparing risk with other risk; there is more to life than the avoidance of risk. Less risk doesn't automatically mean a higher quality of life. Where the risk associated with the generation of electricity is compared with the risk of contracting lung cancer from smoking, the argument is even less defensible. They truly have nothing to do with each other.

Still another approach to HSISE is cost-benefit analysis, the most analytic of the tools, and the one most favored by risk analysts. You simply do the best you can to quantify the risk of the technology under discussion—*all* risk, including beautiful sunsets and lives—and then do the same with the benefits, again *all* the benefits. Whichever prevails determines the best decision. This was the intent of the Executive Order of 1981. Unfortunately it is extremely hard to implement, and people tend to resent impersonal analysis of matters on which they have personal feelings. The Pinto gas tanks are a case in point.

One of the great advantages of cost-benefit analysis is that the costs of safety can't be ignored. That it is hard to do the analysis well is irrelevant to the underlying rationality—something worth doing is worth doing badly.

Because much of the cost of enhanced safety is borne by large impersonal organizations—governments, corporations, whole industries—it is easy to forget that the resources must come out of our society's coffers. (Remember the zero-sum game.) When we pay for something openly we tend to be a bit more critical about what we are getting for our money, but conversely we learn in childhood never to look a gift horse in the mouth. Those who act as if risk reduction were free are guilty of practicing the cornucopia theory of risk management, which we mentioned in Chapter 6—as long as the payers are sufficiently anonymous they can be assumed not to exist. But we are all, in the end, the payers, and we should be sure we are getting our money's worth out of even invisible taxes. Cost-benefit analysis, however imperfectly practiced, is a way of doing this.

This is not just a matter of saving money. While a society is not exactly a zero-sum game, competition for resources is real, and assets squandered on ineffective or inappropriate risk reduction must come from somewhere. Some of the things we now do to reduce risk to the absolute minimum have the opposite effect, while wasting our substance. In our eagerness to save every last life in the next century, we are sacrificing the lives of our friends and neighbors now. Not intentionally, of course, but just as effectively as if it were.

Once the HSISE question has been decided, a question remains: to use risk assessment as a tool for risk reduction through regulation, or as a tool for the evaluation of the effectiveness of regulation. These are different. In the first instance each regulatory agency determination, or each societal decision, would be subject to an assessment and an HSISE test before its implementation. In the second usage decisions would be made according to other criteria, and the role of analysis would be only to judge whether the other criteria are adequate, i.e., pass the HSISE test. If there were no uncertainty in the analysis the first use would be reasonable, providing an impartial technique for decision making.

But there is inevitable uncertainty, and we are not automatons, so in fact we make our regulatory decisions on the basis of all the information we have, not all of it objective and quantitative. Then risk assessment can play a different role, answering the question of whether this less systematic form of risk management is doing its job well enough to satisfy the HSISE determination. This is a matter then of assessing the health of the patient, rather than supporting the diagnosis and treatment of the disease, and is better suited to the current state of the art of risk assessment.

Unfortunately, as was said at the beginning of the chapter, management and assessment tend to be confused. It is the difference between writing a book and criticizing one—both are honorable enterprises, but one is harder than the other, and they require different skills. There are two reasons why many agencies are reluctant to use quantitative risk assessment: one is the concern that the assessment may interfere with the normal decision-making process—that is the confusion mentioned above—the other is a natural, but not commendable, distaste for having their performance evaluated, lest they be found wanting.

The HSISE question has no single agreed answer. Yet a society that values its resources, and recognizes that resources consumed in one enterprise are not available for another, has to pose it and try to answer it. Even if there is no universally accepted answer, especially in a participatory democracy, facing up to the question is an obvious first step toward rationality in risk management. As was said earlier, anything worth doing is worth doing badly—that's how we learn to do it well. How does anyone learn to play the violin?

9

Uncertainty, Causality, Detectability

If one event leads directly to another they are said to be causally connected—a jump to the pavement from a high building will almost certainly lead to an early and messy demise, as will a long game of Russian roulette. Yet a one-shot game of Russian roulette doesn't have a unique outcome, and it is possible to survive it. Similarly, a trip into a mosquito-infested jungle in central Africa won't necessarily lead to a case of malaria, though it might. There is what we call uncertainty in the outcome, where the use of probability is most familiar, and as we have used it up to now.

If we invert the order in which we look at cause and effect, the discussion remains the same. The previous paragraph dealt with uncertainty in the outcome when we know the cause—the cause may also be hard to determine when we know the outcome. The National Transportation Safety Board is responsible for the investigation of transportation accidents (among other duties), and is best known as the organization first on the scene of a major aircraft accident. In the end it produces a report assigning a "probable cause" for the accident, not a "cause." There are usually several contributing causes, and it isn't possible to establish which was *the* cause. This is uncertainty of cause.

The use of the cause-effect relationship in risk analysis involves uncertainty in both directions. A cause may not have a unique effect, and an effect may not have a unique cause. It is important to emphasize that this uncertainty is intrinsic, not just a matter of not having spent enough time and effort to work things out.

This is a problem for the legal system, geared to the idea that the cause of a misfortune can be found through the expenditure of enough effort, and the employment of enough experts. It isn't always so.

The courts have drifted in the last few decades toward what is called strict liability, the assignment of blame (and therefore of liability for damages) to people whose principal sins are to have been in the chain of events that led to the damage and—most important—to have the money to pay. This has come to be known as the "deep pockets" issue—don't seek damages from the party most at fault, extract them from the one who can best afford it. It is a form of income redistribution. Until recently a claim for damages in a civil suit in the courts required that one show that the party being sued was negligent. That is less and less true, because of the powerful need for a culprit. It is argued that the person damaged ought to have *some* compensation, so *someone* ought to pay. One indirect way to acknowledge uncertainty of cause is to sock it to the richest.

Lung cancer is a more subtle case. There is no serious question that smoking has caused the increase in lung cancer in the last few decades. (In men, more than 90 percent of lung cancer is due to smoking cigarettes; in women somewhat less.) Despite that, the courts are still wrestling with the problem of whether any *particular* case of lung cancer is due to smoking, and the invocation of reasonable doubt is winning many court cases for the tobacco companies. No specific case can be attributed to smoking with certainty, though certain types of cancer (squamous-cell and oat-cell carcinoma) are especially well correlated with cigarette smoking. Correlation is not causality. So the cause of the disease is known, but not the cause of any specific case.

A completely separate use of the word "uncertainty"—one that will come up most often in Part II—appears when our needs exceed the state of knowledge about a subject. Dioxins were pernicious trace contaminants in the infamous Agent Orange that was used as an herbicide during the war in Vietnam, but it is not known whether the exposed troops were appreciably damaged. This may sound the same as the uncertainty of cause, but it is not. That had to do with situations in which there were several possible contributors to the event in question, and one couldn't be sure which was most important. These are situations in which the basic scientific knowledge necessary to understand and confidently analyze the case is just not available.

This doesn't mean we are completely ignorant, but only that we are not certain. Some argue vehemently that Agent Orange did damage the health of the exposed soldiers, and they may even be right, but the present state of knowledge doesn't make it possible to say so with confidence.

Uncertainty is not a dirty word. Every scientific measurement and every scientific estimate involves some de-

gree of uncertainty, sometimes small and sometimes large,
but always present. Unfortunately the everyday use of the
word carries with it a sense of confusion, ineptitude, and
inadequacy. The thesaurus that comes with the text editor
being used here offers the following as possible substitutes
for the word "uncertainty": concern, doubt, indecision,
mistrust, skepticism, suspicion, etc., so it should be no surprise that the word carries a negative connotation. In risk
analysis, as in all scientific activities, it is never true that
we know everything, nor is it ever true that we know nothing, and uncertainty is a legitimate and necessary measure
of how sure we are of what we think we know.

Abraham Lincoln said, "There are few things either
wholly good or wholly evil. Almost everything of government policy especially is an inseparable compound of the
two. So that our best judgment of the preponderance between them is continuously demanded."

For all of the cases that follow there are ways to assess
the uncertainty in the estimates, and it will sometimes be
critical to do so. In science, a measurement or estimate
without a statement of its uncertainty is of no use whatever. Probability, and its cousin Uncertainty, are the heart
and soul of an understanding of risk.

Sometimes the range of uncertainty is such that even
the largest reasonable estimate amounts to insignificant
risk, and in that case we might well take a lesson from
the lawyers. A legal expression, *de minimis non curat lex,*
translates into the dictum that the law does not concern
itself with trifles. It means that, even in the exasperatingly
nitpicking business of law, some nits are really and truly
beneath notice; it is meant to protect the legal system
from wasting its time on inconsequential matters. It is
not necessary to decide each case, however trivial.

We, on the other hand, search for risk at the smallest possible level. The search for small risk is of course a
game without end, because modern technology makes it
possible to detect that very risk at levels so low that they
were unthinkable a few decades ago. For some additives
or contaminants of foods or drugs, detection technology
is now so sensitive and so selective that one can detect
only a few molecules of a suspicious substance. Combine that capability with the belief that if, for example,
a large dose of anything can do you in, it must also be
dangerous in smaller amounts, and you have a potential
for mischief. The press will call it deadly, and the hunt for
small traces will be on. Given that psychology, the search
for risk cannot fail to be successful, as was the search for
witches not many centuries ago. In 1988, in response to
an outbreak—almost an epidemic—of campaigns to put
warning labels on all sorts of consumer products, the *Los
Angeles Times* published an editorial suggesting an all-
purpose warning label: "WARNING: THE VERY ACT OF
LIVING IS KNOWN TO BE DANGEROUS TO HEALTH." This
is not entirely funny—we could learn a great deal from
Æsop's fable about the boy who cried wolf. Truth in la-
beling is a mixed blessing. While justified as a matter of
simple honesty, and helpful to those who are particularly
susceptible to the product, it can also dull our awareness.
The label on my microwave oven warns me not to operate it with a damaged door hinge, but doesn't warn me
against all the other dumb things I could do with the machine. Should it? That's quite a list.

Some may think that a weak argument, on the grounds
that a warning label that tells the truth is valuable if *anyone* ever reads it and learns from it. But the theory of
communication, pioneered by Claude Shannon, teaches us

that what is important in communication is not the signal transmitted but the ratio of the signal to the noise. (Noise is interference that doesn't convey information, like the "snow" on a television screen.) Unnecessary labels, like unnecessary regulations, distract us from important matters, and are a net disservice. It is the by now familiar issue of testing not only to see if something is good, but to see if it is good *enough*.

We have said that some regulatory agencies appear to be drifting toward the position that an acceptable risk is one that imposes on the average person a threat of death of less than a chance in a million. For perspective, recall that the average American's chance of being murdered is a hundred times greater, *every year*, which makes a chance in a million seem mighty small.

If a *de minimis* risk is one in a million for an average person, what do we mean by "average person"? Should we be concerned with maximum individual risk—the risk to the most exposed person—or societal risk, the total damage to our society? It makes a big difference. There are defensible ethical arguments that we should strive for the greatest good for the greatest number, just as there are defensible ethical arguments that we are obliged to protect those who are least able to protect themselves.

Given our current death rate of about two million per year from all causes, an additional one in a million would be a total of two deaths per year in the entire population. These would have to be very distinctive deaths to be noticed at all.

Some deaths *are* distinctive. The death of seven astronauts in the Challenger space shuttle disaster in 1986 shook the nation, and effectively grounded the space program for more than two and a half years. If that happens again, and there is no persuasive reason to believe that it won't, the consequences are apt to be longer lasting. When thirty-one people died (mostly fighting the fire) at Chernobyl, the world noticed. We have mentioned that smallpox has been eradicated, so a single case anywhere would be a major event.

But what if the effects we are trying to pin down don't stand out from the crowd? Consider the plight of the anti-nuclear group who tried to establish that the Three Mile Island accident in 1979 caused excessive cancer mortality in the townships "downwind" from the plant. These townships have an overall population of some 25,000, so one would expect about 250 inhabitants to die each year, around 50 of those from some form of cancer. The state of Pennsylvania followed this up by studying the mortality statistics from that area, over the three-year period from 1982 to 1984. During that time the normal cancer mortality would have been about 142.

The anti-nuclear group claimed that there was excess cancer mortality after the accident, and presented in support what later turned out to be biased statistical analysis (according to the Pennsylvania Department of Health). When it was done right the observed number of cancer cases turned out to be 144, compared to the expected 142. Given the square-root-of-N rule, we would expect fluctuations of about twelve in this number, so the difference between 142 and 144 is of no significance whatever. Any increase of less than 20 percent would never have been observed. The claim suffered from other defects, but failed for statistical reasons alone.

To say that a risk is beneath notice means that there is some standard for an acceptable level of risk. It need not be quantitative, it need not be clearly stated, and reasonable people may even disagree about it, but it must exist. Even those who by their actions (and lawsuits) in specific cases behave as if there is no acceptable level of risk, will generally concede the point in the abstract.

Yet it is easy to flounder on the specifics. Some food color additives we will discuss in Chapter 12 are banned because they pose a lifetime risk of less than one in a billion, and even that is an exaggeration. To act against such trivial threats is utterly irrational. An egregious example of an overblown risk that has paralyzed our government for years is that associated with the storage of high-level radioactive waste. This author has rarely read an article or heard a speech or listened to testimony in Congress or seen a television program that didn't portray this risk as truly awesome, and the problem of long-term safe storage as technically beyond our capability. Yet the best estimates of the lifetime risk to the average citizen lie in the range of *one in a thousand billion*. If that can't be called negligible, nothing can.

Under other circumstances it is considered a sign of mental health to accept reasonable risk; unreasonable fears (even when there is a certain level of real risk) are classed as mental illnesses, phobias. A phobia, according to Webster, is an irrational and persistent fear—note the word "irrational." An acrophobic is a person who is unreasonably terrified by heights, even though there may be real risk. The distinctive feature of a phobia is exaggeration of the risk well beyond the point of reality—big bucks are spent to cure it. Not so for the societal exaggeration of risk, though one distinguished Washington psychiatrist has indeed written of nuclear phobia, recommending that it receive the distinction of being treated as an identifiable disease.

There is no good solution to the problem of risk exaggeration, except through a better-informed public and body politic. This must in the end lead to better risk regulation, and to the establishment of either explicit or *de facto* standards of *de minimis* risk wherever it is possible. There are many opportunities—we have mentioned nuclear waste disposal and some food colorings, and others will appear in Part II. When we waste our time and resources on small risks, and at the same time paralyze ourselves, we are behaving precisely as a phobic society.

We can illustrate an extreme *de minimis* issue through a real example, the harmful effects of radiation, which will be elaborated in Chapter 15. For this discussion all that matters is that we don't know whether small exposures are bad, and want to ask whether it is possible to find out by direct measurement.

We could look at the statistics of people exposed to medical x-rays, to see whether a disproportionate number of them suffer ill effects. For typical medical irradiation this kind of blind search is statistically hopeless, for two reasons. One is the size of the exposures, which are so small that the best current estimates are in the range of a chance in a hundred thousand per exposure of developing a radiation-induced cancer at some later time. Even if it were ten times larger, it would have to be measured against a background of "natural" cancer, which kills 22 percent of us, and afflicts even more.

Even if the search were for an effect of one in ten thousand, ten times the expectation, against a background of one cancer death in five, how would it be done? Let's work that one out in detail. First we'd start with a sample of at least ten thousand people, or we wouldn't expect to see even one case. But of those ten thousand people some two thousand would be expected to die of cancer anyway (barring the discovery of a cure), with statistical fluctuations the square root of two thousand, or forty-five cases. Even if we knew accurately how many cases to expect from other causes, and people didn't move around, and people's habits and diet stayed constant long enough to collect the data, the one possible radiation-induced case would still be invisible against the statistical fluctuations in the normal incidence.

A sample of twenty million willing subjects would be needed to make this work. Then, if the radiation risk were one in ten thousand, we'd have two thousand radiation cases, the normal cases would be four million, the square root of that would be two thousand, and we would barely break even in terms of detectability. If something is barely detectable, no one will believe it. Only a study involving the entire population of the United States would make the effects credible, even with this overestimate of the risk. Such effects simply cannot be detected through statistical studies.

This kind of statistical problem afflicted the reporting of the Chernobyl accident, where the predicted late casualties will never be observed. That doesn't mean they will never occur, only that they would be hidden in the normal fluctuations of "natural" cancer. Of course these exposed people will have their health monitored for decades, so that any effects that do appear will be recognized.

What about indirect evidence, from other forms of radiation? For example, cosmic rays are a non-medical source of radiation exposure. They come from the skies, and are unavoidable. The average annual radiation exposure from this source, at sea level, is more than that from a typical chest x-ray, which is in turn less than a fiftieth of that for an x-ray study of the upper gastro-intestinal tract.

In Denver, the mile-high city, the atmosphere isn't as good a shield as it is at sea level, and the cosmic-ray dosage to the inhabitants is nearly twice as high. Even the ground in Denver is more radioactive than the national average. Should we expect to see more cancer in Denver's population of about 1.5 million? The best current estimates are that an additional two or three radiation-induced cancer deaths per year would be expected, because of the high altitude and hot ground. These would be invisible compared to the thousands of normal cases, even if there were a comparable sea-level city to compare with.

The same statistical problems afflict other efforts to measure small risks. When we fall back on animal experiments, in which the environment can be more effectively controlled, large numbers of animals are still required. People speak of "megamouse" experiments, and in fact experiments with tens of thousands of mice have been performed, but that is still not nearly enough to get at very small risks. More about this in Part II.

We devote a great deal of time and effort to risks too small, as a matter of principle, to be observed. Some of them are so small compared to life's normal hazards that they are candidates for benign neglect (to steal a term from Senator Moynihan), and some so small that we can't even be sure they exist. It would be sensible to ignore them, and get on with the real business of life.

10

The Delusion of Conservatism

During their formative training, engineers are programmed with an irresistible drive to design systems and structures conservatively. The profession radiates evidence of this. The tables that tell what to assume for the strength of steel give numbers far below what a steel beam will actually hold before breaking. The building codes provide standards and criteria so conservative that one could easily remove every other stud, or more, from the walls of a typical house, with no visible effect on its structural integrity. A suspension bridge could lose half its vertical support cables and not collapse (the Tacoma Narrows Bridge absorbed extraordinary punishment, far more than its design basis, before it finally gave up the ghost). And who has not crowded thirteen people onto an elevator licensed for twelve, and lived to tell about it?

Conservatism in design and specifications isn't a compulsion to waste effort and materials. It is the accumulation of centuries of experience that the conditions of the real world aren't always predictable, and that it makes good sense to provide some margin for error or for unforeseen events. In any large project mistakes will be made. Wrong keys will be pressed on a computer, workmen will show up for work showing the effects of overindulgence, pieces of wood or steel or other materials will be of lower than expected quality, people will steal, cheat, and malinger, as well as make honest mistakes, wood will rot, steel will rust, and so forth. A design engineer would be foolhardy to believe that what appears in the drawings and plans will exactly describe the finished product, or that it will always be used within its design conditions. He would be equally foolish to believe that he himself is perfect. Those are the bases for conservatism.

Even so, buildings collapse, bridges fall into rivers, dams rupture, and airplanes crash from structural failures. We have learned from experience to do things with just enough conservatism to provide assurance that it doesn't happen too often, and "how safe is safe enough" is determined by common consent, not by any fancy analysis. Adequate safety is when things don't break too often. It may not be elegant, but it works.

But conservatism can't always be used to make allowance for contingencies; it is a luxury. The extra material required to build a bridge five or ten times stronger than necessary is not only expensive, it is also heavy. For some projects the conservative approach carries with it penalties that can't be sustained or ignored. An airplane built to bridge standards of conservatism would be extremely safe, so heavy that it couldn't get off the ground. It wouldn't be of much use either. The moral of a Thurber fable is that you shouldn't lean over too far backward to avoid falling flat on your face. Airplanes are therefore designed and built with much smaller safety margins than bridges—the wings of commercial aircraft typically have a safety margin of about 50 percent, meaning that they are meant to be about 50 percent stronger than needed. Yet the wings of an aircraft rarely suffer severe structural damage in flight, so it works. The safety margins on the structure of the space shuttle are smaller still, because of the even more pressing need to minimize the dead weight. Those margins are shaved to a gnat's whisker.

Audiences who hear for the first time of the small safety margin for commercial aircraft wings are horrified, and often ask whether the safety margin couldn't be increased, say, to a factor of two. The answer is of course yes, but it would unfortunately make the airplane less safe, rather than safer. In return for the illusion of safety provided by stronger wings (illusion because, in fact, wings seldom fail) the airplane would be made heavier and less maneuverable, could carry less fuel, would have more challenging flight characteristics (most accidents are caused by pilot error, not mechanical breakdown), and would be a worse airplane. So the peace of mind afforded the passengers by the strong wings would be bought at a high safety price.

This example illustrates an important principle, the real subject of this chapter: conservative actions are not guaranteed to produce conservative results for complicated systems. A corollary is that a complicated system cannot be made safer by focusing attention on one isolated part of it, and then making that one part stronger or more reliable. It may do more harm than good. Hilaire Belloc said, "Always keep a-hold of nurse, for fear of finding something worse." What engineers say today is, "If it ain't broke, don't fix it." An airplane that has been repaired is both stronger and heavier than it was when new. The strength affects the one part that has been repaired, but the heaviness affects the whole airplane. Safety assurance requires a holistic approach.

A conservatism introduced into the design of something is a deliberate error, albeit in a direction that is expected to be salubrious. When a designer assumes that the strength of steel is 18,000 pounds per square inch (a common design strength), knowing that it is a rare piece of steel that will fail much below 100,000, it is a little white lie and a game of pretend. To be sure, the intent is honorable, and it is hard to imagine how underestimating the strength of materials can possibly weaken the design. The reasoning is sound in most cases, but not all.

There is a wonderful theorem in formal logic which is so subtle that most people don't believe it when they first hear it. Here is a short course in logic, as academic logicians view it.

A logical system is a set of statements, called propositions, a set of starting propositions, called axioms, and a set of rules of inference, which are rules for "proving" one statement from others. The best-known rule of inference is that of syllogism. A syllogism could take the following form: all cats purr when stroked on the tummy; Felix is a cat; therefore Felix will purr when stroked on the tummy. The propositions are in strict order—many of us purr when stroked on the tummy, but are not cats. (A failure to adhere to the proper order leads to what is called the fallacy of the excluded middle. It is common.) There are variations on the rules, and there are other rules, so that, given a set of axioms or starting propositions, it is possible to explore which new ones can be formed according to the rules. When that is possible, it is called proving a theorem.

Now, skipping the subtleties that go with the definition of "truth," we can ask such questions as whether all true propositions can be proved, whether all false ones can be proved false, whether any false ones can be proved true, whether there are any propositions that cannot be decided one way or the other, and whether, in effect, the whole system is self-consistent.

Over fifty years ago, the mathematical world learned (the famous theorem of Gödel) that there are propositions or theorems in mathematics which are undecidable—cannot be proved or disproved, though in a larger sense they may be either true or false. What is important here is a closely related result that states that if the axioms of the system are inconsistent with each other, or at a higher level, if one of them is false, it is then possible to prove any proposition, true or false. This is not obvious. A simple mathematical illustration might be that we assume, as an axiom, that one is equal to zero. (We know that to be false, for other reasons, but assume it to be true for logic testing.) It is then easy to see that, by suitable addition, subtraction, multiplication, and division, we can make any arithmetical statement we wish, true or false. This means we can invent a new formal arithmetic, just like the usual except for the addition of the one axiom, and then all bets are off.

This ties into engineering conservatism because a conservative assumption about anything is a deliberate falsity, and can therefore have any consequences, good or bad. It is therefore wrong, simply and logically wrong, to assume that conservative assumptions—deliberate misstatements of fact—will always lead to a conservative outcome. This is an abstract truth with real consequences, and we have stated it in a formal way just to emphasize the generality.

The strength of airplane wings was one real-life example; there are many others. During one famous fire at the Browns Ferry nuclear power plant, the managers refused to use water to put out the fire, for fear of electrical problems (they were conservative, and conservative fire-management doctrine is to not use water on electrical fires). By holding back they allowed the fire to get out of control, and far more damage occurred before water was finally used to put it out. Buildings that are rigidly reinforced to have adequate strength to deal with earthquakes will fail where more flexible ones won't. Heavily armored knights often did badly in warfare in medieval times. Concern about the risks of vaccines is eating into the immunization level of our population against a number of diseases, and leading to sporadic epidemics of preventable illness. Measles is on the rise in our inner cities. Excessive testing of vital emergency diesel generators at nuclear power plants, to assure their availability in time of need, is wearing them out. If the airlines didn't fly in bad weather, more people would drive, which is riskier. Reluctance to accept medical x-rays, for fear of radiation, can leave serious diseases undiagnosed. And so forth.

These are straightforward cases, but some are less obvious. Underestimating the strength of a material makes it possible to miss the fact that it may hold together long enough under duress to put a more serious stress on a more vital component. If a roof in Southern California is strong enough to accommodate a heavy snow load (surely a foolish conservatism), it may do more serious damage when it collapses during an earthquake. If astronauts (another real case) are forced to practice for all conceivable emergencies, they will be less well prepared for those few that are more likely.

Finally there are conservatisms in calculations, a bit harder to exemplify. A deliberate error in a calculation, in a direction intended to be conservative, provides no guarantee that the result will be conservative. Yet it is widely believed to be true. When seeking a license for a nuclear power plant an applicant must prove that the plant can survive a major loss-of-coolant accident, and the rules specify in detail what deliberate errors (a.k.a. conservatisms) must be made in the calculations.

None of this implies that conservatism is out of place in engineering design. The chapter began by recognizing that conservatism is meant to protect against the unforeseen. And one of the few things that can be foreseen with confidence is that the unforeseen will occur. But any conservative design should, in the end, be looked at as realistically as possible, taking into account such things as the real behavior of materials, to see whether additional risk in the guise of conservatism has been inadvertently introduced. It is much more difficult, and there are few engineers with the necessary skill. Any engineer will concede that it is easier to determine when something will not break than when it will.

In the end, the analysis of the performance of a system, the assessment, should be carried out without conservatism, in short truthfully, if the purpose is to really know how the system will behave under stress. The engineer should put on a different hat, so to speak, when he shifts to an analytic role. Many engineers find that hard—a lifetime of training in conservatism works against it. If the design is done conservatively and the analysis with realistic assumptions, one will have the best of both worlds, and may even discover that what appeared to be a conservative design really wasn't. There is an element of cheating in using the same tools for design as are then used to judge the adequacy of the design, just as it is wrong to grade your own exams in school. Conservative analysis should not be used to check conservative design.

Chapter 15 will contain a particularly good example of the errors of overzealous conservatism—the response to a commercial nuclear accident—where the apparently conservative response is to evacuate everyone in sight at the first hint that radiation may be released. The great majority of accidents will never develop into full-fledged catastrophes, so the belief that the emergency-management authorities should always act as if they will can only lead to trouble. People don't jump out of skyscraper windows into fire nets every time there is a wastebasket fire.

By the same argument, we should not prepare for all possible wars as if they were destined to be nuclear cataclysms. Almost certainly, our military preoccupation with the possibility of nuclear war accounts in part for the below-par performance of the armed services in recent years, when they have been called upon to do smaller jobs. The worst-case fixation leads to failure in the better-than-worst world.

Besides, finding the worst case in any realistic scenario is even more difficult than becoming the best oboist in the world. We can play a game, in which we agree on an accident situation and you tell me your version of the worst case—I can always invent a worse one. After all, I don't have to be reasonable, only worse. After the accident at Three Mile Island, the author's friends often agreed that not much harm was done (except to the reactor), but were still worried, and asked, "What is the worst thing that could have happened?" The standard answer soon became, "The accident could really have gotten out of control, and the containment of the reactor could have blown up, just as a tornado appeared over the area. The tornado could then have sucked up all the radioactivity inside the containment, and traveled through the East, dropping just enough on each large city to render it uninhabitable. And if that isn't bad enough for you, I can think of something worse."

This was obviously (and deliberately) an unsatisfactory answer, but it served to make the point—there is never such a thing as a worst case. If you get close to it, you are speaking of events so improbable as to not be worth considering. Friends often answered, "Oh, come on, be reasonable. Tell me something that had a real chance of happening." When they asked about a "real chance" they were talking probability, and the circle was complete. Risk without probability is, as the French say, like a meal without wine or a day without sunshine. Yet worst-case planning pervades our society. As this is written, NASA is struggling, not very successfully, to wean itself of its dependence on worst-case analysis for shuttle safety assurance, and it is not the only offender. There is an almost hypnotic appeal to thinking that if you have covered the worst case, you have covered everything. But it isn't true.

Part 2

Specifics

11
Toxic Chemicals

FAMILY CHARACTERISTICS

Chemicals, like most of the products of technology, can confer great benefits or can do damage. Some chemicals can cause cancer (those are reserved for the next chapter), and some can poison us in other ways. When they do cause cancer the effects are usually cumulative and progressive—the effects of small doses add up over the years until the disease finally appears. There is usually no evidence of a threshold—a dose below which the chemical is harmless—and the statistics of small effects make it unlikely that we would ever find one, even if it existed.

By contrast, most toxic chemicals that do not produce cancer *do* have thresholds for harm—a large enough exposure can do great damage, while a sufficiently small exposure is benign, or can even be beneficial. Most chemicals are toxic at some dose, and overexposure can lead to damage. The important issues of safety have to do with how and to how much we are exposed. (When asked to define a poison, one chemist said that a poison is too much.) Aspirin, for example, is a known toxic chemical, for many years one of the more common sources of poisoning of young children in the United States. There are now packaging laws designed to meet this threat—no more than thirty-six baby-size aspirin tablets (one-fourth of adult size) may be sold in a single bottle—but aspirin poisoning, though far less common than it once was, is still a perceived problem. It is not much of one, largely because aspirin sales have given way to the sales of the much more highly promoted (and profitable) combination analgesics; in fact, no children under the age of fourteen died of aspirin poisoning in 1985. Many people take aspirin for their whole lives, consuming many thousands of times the lethal dose, with little apparent additive effect, though gastric irritation is common. The body is able to repair any immediate damage, and stay ahead of the game. We have learned to take aspirin in small doses; few are damaged, and even fewer die.

In the proper doses, aspirin is not only benign, but is one of the most versatile and safe medicines known to man. It has been in medical use for nearly a hundred years, and some 80 million tablets are consumed in the United States each day. There is even a growing consensus in the medical profession that healthy adult males should take aspirin on a regular basis, for the protection it provides against heart attacks. Virtually all medicines—probably all—have this characteristic that they are beneficial in the proper dosage, but dangerous in larger amounts. One of the figures of merit for a medicine is the ratio between the lethal dose and the therapeutic dose, the larger the ratio the better.

We won't be talking of medicines here, but of toxic

chemicals which may not have a known beneficial medical effect, but still have a threshold for damage. One shouldn't be too dogmatic in these matters—new uses sometimes turn up for old chemicals. The problem with toxic chemicals is to avoid overexposure, not exposure per se, and that makes regulation easier.

There are two exceptions to this normal pattern. Some toxic chemicals are not easily excreted or chemically transformed by the body, and can therefore accumulate in the tissues, organs, and bones. The difference from carcinogenic behavior is that the substances accumulate, not the effects. The best-known examples of substance accumulation are those of the heavy metals like lead and mercury, where repeated exposure can lead to a buildup of the body stores of the materials, with consequent long-term damage. Lead poisoning from lead-lined wine jugs and water pipes, and from a special condensed grape juice called sapa, which was simmered in lead-lined pots, has been implicated by some historians in the decline and fall of the Roman Empire. Mercury poisoning was the source of the expression "mad as a hatter," and presumably afflicted the Mad Hatter in *Alice in Wonderland*, though Alice never told us that explicitly. (Mercury is used in the manufacture of fur felt hats.) At this writing the Nuclear Regulatory Commission, guardian of commercial nuclear safety in the United States, is prohibited from using the water fountains in its spanking new building, because lead-based solder was used in the plumbing. Let them drink wine, as Marie Antoinette might have said, but not from lead-lined jugs.

The other exception to the threshold rule occurs when repeated exposure to a toxic chemical, in "safe" doses, produces biochemical changes in the body, which then accumulate with time into real damage. Two well-known examples are alcoholism and drug addiction, whose destructive potential needs no elaboration here.

Similarly, chronic bronchitis and emphysema, scourges principally attributable to smoking, do their damage to the lungs over a lifetime. Residents of large industrialized areas live their lives in a pall of atmospheric pollutants, which exact their health toll over decades. Chemicals can exhibit a cumulative effect on the body, even without cancer, but that is not the usual situation. Even so, some types of cancer are caused by constant irritation of cells, whether by toxic chemicals or by physical means. None of these distinctions are perfect.

For any given case, it is usually possible to determine the threshold level for damage, and to find ways to limit the exposures accordingly. It may not be easy, and the results may not be precise, but it can be done. But there are too many cases.

We know of about five million chemicals, of which we use approximately 65,000 in industry in the United States. Each poses a potential toxicity problem, and in fact one-fifth of the chemicals are produced in quantities over a million pounds per year. The vast majority have never been adequately tested for their damage potential. They are in our food, air, and water, sometimes intentionally and openly added, sometimes inadvertently added, and sometimes otherwise added. They are sometimes released to the environment in massive accidents—it was chlorine at Mississauga and methyl isocyanate at Bhopal. They are in our waste dumps, where some are biodegradable by natural phenomena into less-threatening substances, some leak slowly into the biosphere, and others can simply last for many years or even millennia. (The waste dumps of

the past are the favored stamping grounds of archæologists
and anthropologists, and provide much of our knowledge
of early man. That will also, alas, be true of us.) The
wastes must, in the end, go somewhere.

Of course there are laws to regulate the use and disposal of toxic chemicals, and agencies to enforce them—
agencies most often organized along functional lines. The
Department of Transportation (DOT) protects us against
spills of toxic chemicals transported along our public highways and airways, though many of the standards are developed by the Environmental Protection Agency (EPA). The
atmosphere in the workplace is generally the domain of the
Occupational Safety and Health Administration (OSHA),
while the Food and Drug Administration (FDA) and the
Department of Agriculture have various responsibilities
under the Food, Drug, and Cosmetic Act, the Clean Air
Act, and the Federal Insecticide, Fungicide and Rodenticide Act (FIFRA). The Federal Water Pollution Control
Act (FWPCA) empowers the EPA to set standards, but
provides for state and local roles.

There are even constitutional questions about the authority of the federal government in some of these matters. The FWPCA, for example, refers to "waters of the
United States," which has been interpreted by EPA to
include navigable waters, tributaries of navigable waters,
and interstate waters, all of which seems clear enough.
However, the interpretation also includes intrastate lakes,
rivers, and streams that are either used by interstate travelers for recreation or other purposes, are sources of fish or
shellfish sold in interstate commerce, or are used for commercial purposes by industries engaged in interstate commerce. The definition of "waters of the United States" is
so contrived that few exceptions exist. If a Californian like

the author travels to Wisconsin to fish in a stream, that brings it under federal jurisdiction. So much for states' rights.

For atmospheric pollution the constitutional situation is simpler, since no one can argue that the transport of air is not interstate. Indeed, air also fails to honor international boundaries, so the transport of the pollutants that turn fresh clean water from the heavens into acid rain has generated a great deal of tension between Canada and the United States. We'll take this up later.

The various laws that deal with toxic chemicals vary in their definitions of toxicity, and are generally so non-specific that much of the detailed setting of standards has ended up in the courts. The Clean Air Act, for example, specifies as a criterion for regulatory attention that a pollutant "may reasonably be anticipated to endanger" the public health or welfare. The word "reasonable" is much loved by lawyers, for obvious reasons—it provides gainful employment and the exhilaration of combat. Webster says that reasonable means "having the faculty of reason; rational," which is not much help. In the end, the word gets defined by case law.

Some of these issues are exemplified by two quite different chemicals. One, methyl isocyanate, a toxic chemical in wide commercial use, caused the disaster of 1984 in Bhopal, India. For counterpoint, the other is lead, a long-acting and cumulative poison that is widespread in our environment. It rarely kills, but can lead to serious physical and mental damage.

Bhopal—Methyl Isocyanate

Bhopal, a modest-sized city by Indian standards, has a population of about 750,000 (estimates vary). It is situated not far from the geographic center of India, about 350 miles south of the capital of New Delhi, and is itself the capital of Madhya Pradhesh state. Until 1956 Bhopal had a state of its own, which used to be one of the larger Muslim communities in India. For eighty years, ending in 1926, it had female rulers.

Methyl isocyanate is a reasonably common industrial chemical, almost exclusively used as an intermediate material in the production of pesticides, with a production rate of about thirty million pounds per year. Even at that, it is one of the less used chemicals of the isocyanate family, whose total worldwide production adds up to billions of pounds per year. It is extremely volatile and toxic— the NIOSH handbook limits workplace exposure to only 1 percent that of cyanide gas, so methyl isocyanate is considered a hundred times more dangerous. Phosgene, one of the poison gases of World War I, is less toxic, and was used in Bhopal in an earlier step in the production of methyl isocyanate. As one would suspect from its instability in the presence of water, there is no evidence that methyl isocyanate is carcinogenic. It is in fact such an unpleasant material to work with that there were few serious toxicity studies before the accident. Since the other compounds in the family are both more toxic and made in larger quantities, they have attracted most of the attention given to isocyanate toxicity questions.

In moderately low concentrations methyl isocyanate irritates the eyes and respiratory tract, causing temporary breathing problems. Respiratory irritation is a familiar experience to people who work with it, and was familiar to the people at Bhopal. In larger concentrations methyl isocyanate can cause sufficient lung damage to, in effect, suffocate its victims, and can leave survivors with permanent breathing impairment, and in some cases blindness. It is a disagreeable material.

Bhopal and the toxicity of methyl isocyanate came together Sunday, December 3, 1984, at one o'clock in the morning. The event began a few hours earlier; water was the culprit.

Unlike the other isocyanates, methyl isocyanate is extremely volatile—it vaporizes easily. In fact, it boils at a temperature just barely over 100°F, so it must be kept cool. Heat will make it boil or, in a closed vessel, build up pressure as in a pressure cooker. This can eventually lead to rupture of the container and release of the material. An essential part of the safety prescription for this chemical therefore consists of keeping it cool, while monitoring its pressure. Yet the air-conditioning system normally used for this purpose at Bhopal was shut down at the time of the accident, and had been out of service for many months; it was winter in India, and room temperature seemed cool enough.

Unfortunately (or fortunately for chemical processing) methyl isocyanate is also very active chemically, which means unstable. (Stability is good or bad, according to the needs of the user. If nitroglycerine were not unstable, dynamite could not have been invented.) Methyl isocyanate will react violently with water, and will even react with itself in the presence of certain other materials, those that can function as catalysts. Iron is one such material. (A catalyst is a chemical that can speed up a chemical reaction, without itself being used up. It is an expediter.) In addition, all these chemical phenomena are faster at higher temperatures.

These reactions are also what chemists would call exothermic—the reaction itself releases heat—so the chemical activity can warm the reacting materials, speeding up the reaction and creating more heat, etc. That leads to a form of instability, because the reaction can feed on itself, or at least on its own heat. (That's the way a fire works, which is why you have to light it before it becomes self-sustaining. It's hard to build a small fire, but easy to build a large one.) And higher temperatures lead to higher pressures, for an enclosed tank. There were of course pressure gauges on the tanks at Bhopal, and they were read, but they were not taken seriously until too late.

Those are the precursors of the accident: inoperative, disabled, or ignored safety and alarm systems; a material subject to exothermic reactions, especially with water; an operating crew that had noticed a leak, and had decided to do something about it after the next tea break; a generally rundown plant with leaky valves and a lagging maintenance program; and a highly toxic gas in two of the tanks. Only a trigger was needed, and that was provided in the form of a thousand or so gallons of water supplied Saturday night to the doomed tank of methyl isocyanate. Whether it was done intentionally or inadvertently, it was a mistake. The reaction began slowly—the initial temperature of the tank was around room temperature—and as the water reacted with the methyl isocyanate the pressure rose from about 2 psi to 10 psi. This happened over the first hour. As the mixture warmed the reaction speeded up and the pressure ultimately rose to about 50 psi, at which point the relief systems opened and the poison gases escaped and moved toward the city. Had the early rise in pressure been taken seriously, few of us would have heard of Bhopal.

In the end, about twenty-five hundred people were killed and tens of thousands injured. Since methyl isocyanate gas is heavier than air, it stayed near the ground as it progressed through the countryside. Many of the killed were asleep, and were simply unable to escape. No alarm sounded, allegedly because so many of the workers were busy trying to save themselves or their families.

Bhopal was an appalling disaster, the worst industrial accident in history. Only wars, earthquakes, floods, and the like can kill and injure such a large number of people. Even the *Titanic* carried fewer to their deaths. Of course we knew that toxic chemicals can kill, but they do so so rarely that the world notices when it happens; vigilance is therefore hard to maintain. Scapegoats abound in this case, but we should always remember that the scapegoats in the Bible were allowed to escape, hence the name. The supervisor who took tea, the worker who ignored the pressure gauge, the worker holding the hose that injected the unwelcome water—all were guilty of laxity. Yet when so many things are wrong, the root cause is management. Management problems of this sort are commonplace, bred by complacency. In fact, the chemical industry is one of the safest of our large enterprises, with an accident rate well below the rest of American industry, in part because it has learned from hard experience that chemicals can be dangerous. Those who forget that are soon reminded that nature's friendliness conceals an unforgiving streak. The necessary sense of caution seems not to have been transmitted to India; the risk was technological, but the failures human. The solution is not to banish toxic chemicals, any more than it makes sense to throw away the hammer after we have bashed a thumb with it.

When the chairman of the board of Union Carbide (the parent company) arrived in India just after the accident, he was arrested by the Indian government, and charged with negligence and a variety of other sins. That was not particularly helpful, especially since his motives in making the trip were solid and compassionate. He deserved credit for his prompt response, but the Indian instincts in assessing the ultimate responsibility were generally right. Sadly, vengefulness after an accident is no substitute for vigilance before it.

One of the imperatives of the risk business is to learn from experience. As Santayana said, "Those who cannot remember the past are condemned to repeat it." While it is not true that every accident possible has long ago occurred, there are often underlying similarities among major accidents, and these can cross technological lines.

We take advantage of this in a number of areas, as in the efforts of the National Transportation Safety Board to learn enough from transportation accidents to make repetition less likely. So, we may ask: who had the responsibility to learn the lessons of the Bhopal accident? It was, after all, the most lethal industrial accident in history, and there must be important lessons here, after the finger-pointing ends. Soon after the accident this author phoned a friend in a high position in the White House, to ask just which government agency had the responsibility to collect the Bhopal lessons, especially those that cross institutional lines. His response was, "That's a very interesting question. I guess nobody." And that's the way it stands today, which is why we appoint a new presidential commission every time there is a new disaster. That we need to learn from experience is a real lesson, waiting patiently to be learned from experience.

Lead, Our Decreasing Pollutant

There is plenty of lead in the ground, but not until the

Industrial Revolution did much get into the atmosphere, or into our food or drinking water. Lead is so heavy, more than ten times as dense as water and nearly ten thousand times as dense as air, that any that is swept into the atmosphere by the wind soon falls to the earth. The slight solubility of lead in water guarantees that only stagnant water contains much lead. As a pollutant, the lead in our environment is almost entirely a product of human intervention, and ultimately technology. There isn't much, but it accumulates in the body.

And lead is harmful. Among the symptoms of lead poisoning (plumbism to experts) listed in the *Merck Manual* are personality changes, headache, and abdominal disorders in adults, with seizures, coma, and mental retardation in children. Lead poisoning was in fact the first recognized occupational disease, identified by Hippocrates in the fourth century B.C. Acute lead poisoning can be treated with chelating agents, chemicals that form a chemical bond with the lead, and help the body to rid itself of the intruder. As far as is known, lead does not cause cancer.

Humans have been mining, smelting, and using lead for thousands of years, since before the earliest historical records. Because of its low melting point, it was probably the first metal ever to be smelted. There exist lead coins and lead-lined water pipes from Roman times, to say nothing of the infamous wine jugs. Yet the accumulation of lead in the human body was probably much smaller then than it is today, almost certainly less than 1 percent as much, and we simply don't know whether the substantial lead burden now carried by the average American has population-wide consequences. Residents of the inner cities carry more lead than people in the suburbs and rural areas, but there are other demographic defects of inner-city life that could mask any possible effect of the lead. Children, the most susceptible, have their whole lives ahead to accumulate lead in their bodies. We do know that typical body burdens are not far below the levels known to produce the symptoms described above, so that, even if we are now generally symptom-free, it is not by a wide margin. According to the *Merck Manual* a blood level over 30 in certain irrelevant units—micrograms per deciliter—justifies a diagnosis of moderate lead poisoning. The averages measured during the 1970s in inner-city children were about 20, in the same units, though there is now some question about the reliability of the measurements. And apart from overt symptoms, there is some inconclusive evidence that low levels of lead may affect the IQ of children. The risk posed by lead in the environment is real.

Of the million or so tons of lead used each year, about half is recycled lead, and most of that goes into automobile storage batteries. Although substantially less has been going into the production of leaded gasolines in recent years, the lead that does go into them is directly discharged to the atmosphere, where it is available for human intake. There was more in the past, and it was harmful. Not only did we breathe it, but the fallout found its way into the water supplies and food supply (actually the most important route), so that there was a simple and direct connection between the amount of lead burned in gasoline and the amount carried in our bodies. The atmospheric lead started decreasing with the painful transition to unleaded gasoline, and the body burden is following. Since people do buy new cars to replace old ones, the passenger-car fleet turns over every decade or so, and one might think that all we now have to do is wait for the older cars to vanish from the highways. Unfortunately some antisocial drivers continue to illicitly use leaded gasoline in their cars (it is cheaper), which limits the effectiveness of this lead reduction strategy. In the end, it will probably be necessary to ban leaded gasolines outright, or tax them so heavily that they lose their price advantage.

The reader has doubtless noticed a tone in this book suggesting that we are usually overly panicky about technological risk—the first words of the Introduction say so—but the case of lead is an authentic exception. Lead is a genuine threat, with identifiable sources, doing demonstrable harm to children and adults alike, and we have dragged our feet about doing something about it. Here resistance by the affected industries has played an important role, as has misdirection of effort by some of the so-called environmental organizations, whose management is so hung up with combat against more exciting things that they have little energy left for the humdrum but important.

Nonetheless, something *is* being done about airborne lead. While we haven't delegitimized leaded gasolines outright, we have greatly curtailed their use. The results are already apparent in the atmosphere. The amount of lead annually deposited in the atmosphere by road vehicles has gone down from over a hundred thousand tons in 1975 to less than a thousand tons in 1985 (though other elements of the transportation sector still contribute fifteen thousand tons), and the atmospheric concentration of lead has gone down in the same time from 1.3 to 0.26, in micrograms per cubic meter. EPA has set an air-quality standard of 1.5 in these same units. So there has been real and measurable progress, and in a reasonably short time, but it has not been easy.

Finally, the fact that lead in motor fuels has been emphasized should not be interpreted to mean that that is the only supplier of lead to the human body. Standard plumbing (note the use of the word) practice for many years has been to use copper tubing in new homes, fitted with soldered junctions. Plumbing solder long consisted of about 50 percent lead, some of it in permanent contact with the water in the pipes. Thus the first morning water in a home, after overnight stagnation in the pipe, could well have appreciable levels of lead. It is no longer legal to use lead-bearing solders in domestic water systems, but that affects only the newest homes.

The principal source of lead poisoning in children, especially inner-city children, is lead in the lead-based paints that have been used for many decades. Especially in older homes and apartments, these paints (like all paints) ultimately chip and flake, and are, one way or another, eaten by small children. Even toys were painted with leaded paints before awareness of the hazard became widely disseminated. Largely through the efforts of the Consumer Product Safety Commission, lead-based paints may not now be used on toys, and there is a strict (and small) upper limit on the amount of lead that can be used in any paint.

Control of lead poisoning is one of the success stories for rational environmental management. The threat was real and the necessary measures were taken. Though the costs were and will be high, they are a reasonable match to the benefits. While neither the country nor our bodies have returned to a state of pristine purity, we have moved much further away from the brink, and are going in the right direction.

12

Chemical Carcinogenesis

FAMILY CHARACTERISTICS

There is something particularly fearful about carcinogenic chemicals, and for good reason. The damage can add up silently over a lifetime, with no apparent ill effects until near the end, by which time it is often too late to do much. And death from cancer, chemically induced or not, can be especially unpleasant. Since 22 percent of deaths in the United States nowadays are from cancer of all sorts, it is all too familiar. Finally, exposure to a carcinogenic chemical in the environment may not even be noticeable, and a hidden menace is particularly troubling. Fear of cancer dominates public concern about chemical risk.

Yet, of the approximately 65,000 chemicals in commercial use, over 10,000 of them produced at over a million pounds per year, no more than a few dozen are definitely known to cause cancer in humans. Usually this information comes from studying a cluster of cases, and tracing the cause to the offending chemical. The first discovery that chemicals could cause cancer came from the observation by Percivall Pott in England, in the eighteenth century, that exposure to soot was a reason that chimney sweeps tended to get cancer of the scrotum. Even so, it wasn't until the early twentieth century that coal tar was shown directly to cause cancer in laboratory rats.

In addition to the few dozen chemicals for which there is no reasonable doubt (asbestos, mustard gas, etc.), there are a few hundred chemicals that are suspected of causing cancer with sufficient exposure. Fortunately that exposure is usually far higher than any we normally encounter, and we only know about it because we are willing to subject laboratory animals to enormous doses of the suspected carcinogen. Considerable effort is put into the breeding of special strains of small animals (usually mice and rats) which are particularly susceptible to cancer at certain sites—that makes the experiments easier, but the conclusions harder to interpret. More about this in a moment. For all the effort devoted to finding and identifying carcinogens, carcinogenesis by exposure to chemicals is an exception, not a rule.

The Department of Health and Human Services is required by law to issue an annual report, listing all currently known and suspected carcinogens to which a significant number of people in the United States are exposed. There are currently about 150 chemicals on the list, out of the 65,000 chemicals used in commerce and the five million known chemicals, which is some measure of the seriousness of chemical carcinogenesis. Certainly the widely disseminated image that the country is drowning in a sea of cancer-producing chemicals is far from true. The exception, of course, is cigarettes.

How then do we know whether a suspected chemical

can cause cancer in people? At what dosage and over what time period can the damage occur? Finally, are any or all of us likely to be exposed to enough of the stuff to hurt?

To begin with, it is impossible to avoid the tyranny of the square-root-of-N rule, explained in Chapter 5. When we deal with small numbers and small risks, the inevitable statistical fluctuations in cancer rates can easily lead to erroneous conclusions, in either direction. Typically, the objective of exposure management is to avoid chemicals that have a lifetime chance of not much more than one in a million of producing a lethal cancer in an individual. This means immediately that we are dealing with such small numbers that the statistics are particularly bad, and the "natural" cancer mortality is far higher. Cancer represents a particularly odious risk, so we are willing to go the extra mile to keep it in check.

If we depended on mouse and rat experiments to detect such small risks we would soon deplete the world supply of small rodents. To detect a chance of one in a million, you have to do a million tests, or there will be no effects (on the average) to observe. That's out of the question, so we breed susceptible species and give them massive doses of the suspected carcinogen, just to make something happen. Sometimes, if the chemical is in food and they don't like it, they just won't eat enough. Then it is necessary to fall back on forced feeding, called gavage (accent on the last syllable). In the end, one way or another, and whether they like it or not, the test animals get the desired dosage, and we learn what we seek to know about the suspected chemical.

There is an obvious problem of interpretation. No mouse can eat as much as we of the suspected chemical (no two-pound sirloin steaks for mice), but they do consume *proportionally* far more. What is more, they normally live only a year or two, far less than we. Vast extrapolations are required to interpret rodent data in terms of possible threats to humans, and there are many unknowns.

When a chemical is nominated for carcinogenicity testing, the process begins with a screening test, preferably the Ames test, invented by the noted California biochemist, Bruce Ames. Since cancer is believed to start with an alteration in a gene, the Ames test is not a cancer test at all, but is instead a test of the chemical's ability to cause mutations in a bacterium, typically a strain of *Salmonella.* A carcinogenic chemical will usually, but not always, cause mutations in *Salmonella,* so the mutation test serves as a surrogate for real carcinogenicity testing. By now thousands of Ames tests have been performed on suspected chemical carcinogens, and, unless there is other evidence, only substances that fail the test pass on to small-animal testing. There are other short-term tests, which follow similar principles.

The animal tests are much more elaborate, and correspondingly much more expensive, which automatically means that far fewer are done. So far, only a few hundred chemicals have been tested this way, yet the main problem is not a shortage of tests but the difficulty of interpretation. The number of animals involved in a given test can range up to a thousand, though there have been a few special cases that have used many more, and the duration of the test is limited by the normal lifetimes of mice and rats, about two years. Such a test costs in the neighborhood of a million dollars. On rare occasions, experiments may involve larger animals, but of course fewer of them.

To begin the interpretation question, we truly do not know how to make the body weight and lifetime extrapolations mentioned above. Since the exact causes of cancer remain mysterious, we have to fall back on expert judgment—glorified guesswork. The usual procedure is to scale the dosage by the body weight of the animal, so that a gram of material fed to a mouse (about 20 percent of its total daily diet) may be equated to about five pounds fed to a person (rather more than 20 percent of his daily diet, even for good eaters). There are other ways to do the scaling, and different agencies have different habits, but none is arguably better than another. (Feeding tests have been used here as an example, but the chemical can be administered in other ways. An average mouse weighs about an ounce, and eats about one three-hundredth as much as a human, while a rat may weigh ten times as much as a mouse, and eats about three times as much—appetite doesn't scale with size.) Similarly, continuous dosage over the life of the mouse is usually equated to lifetime dosage for a person, despite the disparity in time. Each of these assumptions has some rationale, but there are nearly as good grounds for alternate approaches. Besides, mice and rats aren't physiologically the same as people.

So there is unavoidable uncertainty in the application of the results of rodent tests to judgments about human susceptibility. Uncertainties of a factor of ten are common, and not scientifically significant in this business, though they can be crucial when the issues get into the courts. We try to err on the safe side when the tests are interpreted, but that leads to overconservatism, with the pitfalls described two chapters back.

The question of the dose-effect relationship is even more difficult than the difference in sizes. We've said that the need to observe an effect requires that the test animals receive large doses of the suspected carcinogen. In effect, we want to learn from large doses in small animals in a short time about small doses in large animals (humans) over a long time. The high probabilities in small animals must be translated into low probabilities in people. That requires what is pretentiously called a mathematical model.

A mathematical model, for these purposes, is a relationship between the magnitude of exposure to the bad stuff and the likelihood of getting cancer from it. The relationship may or may not have some basis in our limited understanding of the process—if it does it is called a theoretical model. If the model is dreamed up out of the whole cloth, but seems to fit the known facts, it is called an empirical model. Both empirical and theoretical types exist. For example, the most used model of the dose-response relationship (perhaps only because it is simplest) is the so-called linear model, which allows one to pretend that the probability of getting cancer from a chemical is directly proportional to the total accumulated dose. Twice the dosage leads to twice the risk, and half as much to half the risk. Another model is called quadratic, which means that the effects are assumed to be proportional to the square of the dose. For this model, twice the exposure leads to four times the risk. (The model most recently favored by the National Academy of Sciences for radiation-induced cancer is a combination of these last two, called the linear-quadratic model.) There are even more complicated models.

Ten years ago the National Center for Toxicological Research performed a famous one-of-a-kind experiment to determine, once and for all, the dose-response relationship of a particular chemical carcinogen, 2-acetylaminofluorene, 2-AAF to its friends. In this monumental test, 24,000 fe-

male mice were used to test the potency of 2-AAF in producing liver and bladder cancers, the two types for which
it is most effective in mice. When the data were in, the
relationship seemed to be strictly linear for liver tumors,
down to the smallest doses used, while it was not at all
linear for bladder tumors, *in the same mice.* So there is no
universally valid dose-effect relationship, applicable to all
cases. Chemical carcinogenesis is sometimes proportional
to the dose, and sometimes not. When it is not, it is almost always smaller at low doses, as in this experiment.
Most probably that is true of other forms of carcinogenesis.
This is important, since all of the testing on small animals
is done at high dosages, and the common assumption of
a linear model means that the risk at the more relevant
lower levels is probably exaggerated.

To see how to do these tests, suppose one starts by
planning to use around five hundred mice—a reasonable
number. Since there may be sex differences (the males
cannot, for example, have cancer of the ovaries), the mice
will usually be divided into two groups. Then, since the
purpose is to learn something about the dose-response relation, their numbers will have to be divided further into
dosage groups. There might be four dosage groups: one
control group, who receive none of the suspected substance; one high-dose group, so something is likely to happen; and two groups in between. That's eight groups all
told, so there would be about sixty animals in each group.
Those numbers are typical.

The problem of extrapolation (really interpolation, to
language purists) is best illustrated through a real case.
That case (an extensive and pioneering formaldehyde inhalation test) involved four exposure groups of about a
hundred mice each, and a somewhat larger number of rats.

Most were exposed to the formaldehyde for two years, in effect their lifetimes. The result, only slightly oversimplified, was that no cancers were observed in any of the three lowest exposure groups of mice (including the group that received no exposure at all), but two cases of cancer were observed in the most exposed group. The rats fared worse, with two cases in the next-to-highest group, and many in the highest. Neither rats nor mice showed any cancers in the two lower exposure groups.

What to make of it? Bear in mind that the highest exposure is far above any we really care about. In fact, the exposures of real interest—leading to a chance in a million, or something like that—lie between zero exposure and the next lowest in this test. No cancers were observed in either mice or rats in either of these exposure groups.

Clearly, there is a problem. Since no cancers afflicted about five hundred animals at an exposure higher than any we will ever see, we might conclude that formaldehyde is harmless. But that would be irresponsible, because a tenfold increase in concentration does seem to induce a substantial probability of contracting cancer (a rare nasal cancer in this case) in one kind of animal, but not in the other. Though only with large doses, and preferentially in one kind of animal, formaldehyde can indeed cause cancer. We have a dilemma. Formaldehyde is currently categorized, for regulatory purposes, as a "probable human carcinogen," though there is no convincing evidence that it causes cancer in people. Note that the formal classification makes no mention of dosage—the classification would be the same if tons of the stuff were needed.

The example was meant to illustrate how uncertain the evidence can be, yet this was a particularly good case, involving more than the usual number of animals. A writer can mislead his readers in such cases by giving the impression that so little is known that we might as well quit. It was noted earlier that the belief that you are either certain of something or know nothing at all about it is wrong; all scientific research falls into the area between these two extremes.

The uncertainties in carcinogenic testing may be somewhat worse than in some other scientific fields, but are no different in kind. When someone claims to be completely sure of something, it should be held against him, but in the public debate over risk the opposite is true. Uncertainty is held against honest scientists who confess that they might conceivably be wrong; an admission of uncertainty is interpreted as a mark of ineptitude. A well-respected former senator once wished for more one-armed scientists, because scientists always say "on the one hand this, and on the other hand that." It is easy to sympathize, since decisions do need to be made; but the uncertainty is real. Not only is there no point in hiding it, it would be dishonest to do so.

It doesn't hurt to put all this into perspective, by asking just how much of our real cancer incidence is due to the man-made chemicals in our environment. Bruce Ames, the biochemist responsible for the Ames test mentioned above, has embarked on what amounts to an educational crusade, aimed at providing just that perspective. It turns out that there are far more natural toxic and carcinogenic substances in the world than there are man-made threats, and that only an extremely small proportion of cancer is due to toxic wastes and pollution. Dried basil contains estragole, comfrey tea contains symphytine, mushrooms contain hydrazines; all are carcinogens. A form of the oft-mentioned aflatoxins appears in the milk of cows who have eaten moldy grain. Ames has estimated that we are taking in natural pesticides in amounts at least ten thousand times the man-made pesticide residues. The reason is obvious. As they evolved, the plants have had to build their own chemical protection against insect pests—we are not the first to have recognized the need. And the plants did not have to get government permission to protect themselves.

The Delaney Clause

In 1958 the Congress confronted the uncertainties in cancer induction in a marvelously straightforward way, by implanting the infamous Delaney clause in the Food, Drug, and Cosmetic Act. This sets a limit of zero, absolutely zero, for the amount of any covered additive that has shown any evidence of causing cancer in man or animals, at any dosage. Such an additive may not be used, even in the smallest amounts. This standard has an appealing simplicity to it, but takes no account of the fact that some substances are extremely carcinogenic, and others hardly at all. Relatively harmless substances are banned, and even some beneficial ones, while far more dangerous materials that are present naturally are unregulated. The only way to comply with the Delaney clause is to show that reasonably diligent efforts to find deleterious effects have failed; that doesn't mean that the additive isn't carcinogenic, only that it has not been *shown* to be carcinogenic. Of course the failure can be dealt with by charging up our diligence and trying harder, as was the case during the search for witches just a couple of centuries ago. (People are occasionally consistent—Congressman James Delaney was an ardent anti-fluoridationist.)

The exact wording of the Delaney clause (for food additives—the wording is slightly different for drugs and cosmetics) is: "No additive will be deemed safe if it is found to induce cancer when ingested by man or animal, or if it is found, after tests which are appropriate for the evaluation of the safety of food additives, to induce cancer in man or animal." One can quibble about the meaning of the words "found" and "induce," but the language has been interpreted to mean an absolute prohibition if there is any evidence whatever of carcinogenicity of the additive. The legislative history supports this interpretation of congressional intent, with the apparent logic that additives are unnecessary, so no risk at all can be justified. It is certainly the most extreme possible position, and at the same time the simplest.

From time to time an effort is made to challenge the burden that the Delaney clause imposes on risk managers. In 1987 the FDA proposed to allow the use of some food dyes—Orange 17 and Red 19—the former of which, though not literally meeting the legal requirement of being entirely without risk, was estimated to pose a lifetime risk of less than one in *nineteen billion* to the average American. That means that one person may die in the next ten thousand years. FDA argued that such a low risk is realistically equivalent to the legal requirement of no risk at all ("the risk is so trivial as to be no risk"), but was turned down by the United States Court of Appeals, and was denied review by the Supreme Court. For orientation, recall that a risk of death of one in nineteen billion is equivalent to the risk of a hundred additional yards of flight on a commercial airline, or ten yards of driving, in a lifetime. The risk is less than the risk of eating one one-hundredth of a peanut with the FDA-approved level of aflatoxins. It is equivalent to spending a few minutes in Denver (with its cosmic rays), instead of at sea level. The Appeals Court agreed that that risk was trivial, but "reluctantly" concluded that it was bound to rule as it did by the clear wording of the law. One feels, in reading the case, that the Court was trying to send a message to the Congress.

The following year EPA announced that it was going to take on the Delaney clause in the case of pesticide residues. EPA, which regulates these under the Federal Insecticide, Fungicide, and Rodenticide Act, announced that it was going to treat an upper-limit lifetime risk of less than a chance in a million as if it were negligible. There are constant stirrings in the federal bureaucracy that chafes under the Delaney clause. Change is in the wind, but who in the Congress will cast a vote that will inevitably be labeled as pro-cancer, and so it stands.

Saccharin

The battle between saccharin and the Delaney clause has been fought to a draw for many years. The special feature of this battle is that everyone knows of saccharin's usefulness as an artificial sweetener, so it can't really be banned quietly. That sets it apart.

A bit over a hundred years ago (1879), saccharin was discovered by accident at the Johns Hopkins University. In the course of some relatively routine experiments on the oxidation of some toluene compounds, one of the chemists noticed that his food tasted unusually sweet; in fact, the sweetness was all over his hands and arms. One of the marks of a good scientist is that he listens when nature tries to tell him something, instead of telling her to shut up because he is too busy. So saccharin was discovered, a substance about four hundred times as sweet as ordinary sugar, and without any redeeming food value, or calories. (Aspartame, the currently preferred artificial sweetener, was discovered in a similar fashion—it pays to lick your fingers.)

Saccharin soon became a fixture in many countries. There were complaints that it was non-nutritive, but safety issues stayed in the background in that early period. Even so, it was briefly banned from foods in the United States just before World War I (partly for safety reasons, and partly because of its lack of food value), but the sugar shortage during and immediately after the war made the ban short-lived.

After World War II saccharin use grew rapidly, peaking at an impressive thirteen pounds sugar equivalent per person per year in 1984, the year that aspartame (about half as sweet as saccharin) came into widespread use. By far the largest portion of the saccharin went into soft drinks and sweeteners for use at the table. Saccharin provides an opportunity to add sweetness to life, both as a substitute for the sugar we might ordinarily consume, and to add even more sweetness to the diet. Judging from the steady growth in the combined sales of saccharin and aspartame, and the 150 pounds of sugar and corn sweeteners we each consume every year, the American sweet tooth is still in the flower of its youth. The per capita consumption of sugar in the United States has gone up by a factor of four in the last hundred years.

By the 1970s the safety questions, always present, could no longer be ignored. Laboratory experiments provided strong evidence that saccharin could produce bladder cancer in small laboratory animals, though there was no substantial evidence that it could do so in humans. In fact, saccharin was, and is, one of the weakest carcinogens known, if not the weakest of all. It is ten million times less carcinogenic than the aflatoxins we have mentioned before. In the society of carcinogenic chemicals, saccharin is a second-class citizen. In fact, the only known carcinogen that is probably weaker is ethyl alcohol, but of course we consume that in much larger quantities as we enjoy our adult potables, and carcinogenesis is not its principal threat to human health.

But saccharin does cause cancer in laboratory animals, so in 1977 the FDA, faithful to the Delaney clause, announced its intent to ban it. Recall that there was then no reasonable substitute to serve as an artificial non-fattening sweetener, so a ban on saccharin would have created dietary chaos, to say nothing of an even flabbier population. (The per capita consumption of saccharin at the time was equivalent to about eight pounds of sugar per year, so direct substitution of sugar for saccharin would have fattened the average American by about five pounds per year. Of course it wouldn't have happened that way.) The potential dietary impact undoubtedly played a major motivating role for the FDA's stalling tactics in requesting study after study during the 1960s and most of the 1970s. Finally, in 1977, FDA had to bite the bullet and act, despite the relative insignificance of the risk.

Did the Congress then see the error of its ways, and repeal or modify the Delaney clause? Heavens no, it immediately enacted an eighteen-month waiver specifically for saccharin, forbidding the FDA from banning the substance while an effort was mounted to find an alternative non-nutritive sweetener. And so it has been ever since. The waiver has been renewed at regular intervals as the search went on, and over the years the alternatives have turned out to be worse than saccharin. Cyclamates had seemed promising for a while, but were completely banned in 1970. At this writing, in 1989, the FDA is reviewing the ban on cyclamates, and may well decide that the original ban was based on inadequate evidence. Cyclamates could stage a comeback.

Finally in 1984 aspartame came along, with health hazards of its own but no known evidence of carcinogenicity. Aspartame is bad for people with phenylketonuria, a genetic metabolic disorder that afflicts about one baby out of fifteen thousand. People with that genetic flaw can simply avoid aspartame, provided the foods containing it are properly labeled, and they are willing to get along without diet soft drinks.

This might have been a good time for the Congress to breathe a collective sigh of relief, because the perceived need to continue the waiver for saccharin was finally gone, and Delaney could once again be enforced without the exception. Instead, the waiver has recently been renewed, for a somewhat longer period, and is effective until 1992. The opera isn't over.

Saccharin has undergone extensive testing in animals, over a long time, in an effort to settle the question of its carcinogenic powers. The testing is still going on, and is of the sort already described, with different types of rodent, and at different dosages. Of course there isn't any controlled testing in people, though the fact that we have consumed so much of the stuff for so long without any obvious surge in bladder cancer means that it can't be very potent. The evidence that cancer may have been induced in humans is ambiguous, and probably not worth discussing here, since the animal experiments are generally considered decisive. Recall, however, the problems of extrapolating from small animals to people.

What recent experiments show is this: no bladder cancer is seen in the subject mice or rats until their lifetime diet consists of approximately 1 percent or 2 percent saccharin, at which point a few percent of the animals develop tumors. From then on, the cancer risk rises with higher concentrations. A 1 percent saccharin diet for mice would be equivalent, in people, to a saccharin consumption of about a quarter of a pound per day. This would in turn correspond in sweetness to about a hundred pounds of sugar a day. One either pities or envies the mice, depending on one's sweet tooth. At the peak, in 1984, our actual average saccharin consumption of about a half ounce per person per year corresponded to about a three-thousandth of that. With the linearity hypothesis, this would suggest a lifetime risk of at most ten chances in a million of getting bladder cancer from consuming saccharin. This was a typical calculation of this type of risk. It wasn't complete, because the other side of the equation was left out, the health effects of the obesity that would surely follow a switch from saccharin to sugar. There are health benefits to saccharin.

One can then either take it or leave it—what follows is no longer science or statistics. For one side the argument is that a lifetime risk of ten chances in a million is irrelevant when more than a fifth of us die of cancer anyway. In any given group of a million people, over two hundred thousand are going to die of cancer, and how could one even tell if there were ten more? Besides, that's equivalent to driving about twenty more miles per year in your car, over your whole lifetime, and why worry about that? The risk is negligible, a tempest in a teapot, and certainly small compared to the risk of sugar-induced obesity. That is one side of the story.

The other side multiplies the small individual risk by the whole population, to see how many people are really at risk. That calculation tells us that if that peak level of saccharin consumption continued forever, *and* the linearity hypothesis is applicable in this case, about ten deaths in every million would be attributable to saccharin-induced cancer. Since about two million of us now die each year, that means we might be unnecessarily killing twenty people per year, perhaps hundreds by the turn of the century. And, so the argument goes, what if you were one of them? That is the other side.

For every small individual risk, one can multiply it by the population of the United States, or even of the world, and show that one is talking about a substantial number of individuals. If we wished to protect each individual in the world, for his or her whole lifetime, we would be speaking of risk levels a million times smaller than those we now regulate. It is not only impossible, but would waste resources that can be put to better use in improving human health. As Frederick the Great said, "He who would defend everything defends nothing."

Our Canadian neighbors, with exactly the same information about artificial sweeteners, have regulated differently. Though the United States has completely banned cyclamates, and the Congress maintains a waiver to keep saccharin legal, the Canadians seem to prefer cyclamates to saccharin. They only distribute saccharin through pharmacies, and do not permit groceries to sell it, while they allow cyclamates to be sold in groceries. They do not permit the use of either as a food additive, so they treat both with suspicion, but in the opposite order.

So, at least for a while, we can go on putting saccharin in our coffee, bearing in mind that the coffee itself, without saccharin, seems to pose a risk of pancreatic cancer. But two things favor the coffee: the evidence against it is not much stronger than against saccharin, and it is a natural product. In fairness, we should say that it is legally possible in the United States for a regulatory agency to ban food containing even a natural ingredient that would "ordinarily render it injurious to health"—a much stronger requirement, but one that could conceivably be met. Still, it would require extraordinary courage, well beyond anything one has a right to expect from a civil servant, to become the government official who tried to ban coffee in the United States. It is bad enough with saccharin. To say nothing of alcohol, which was banned for fifteen contentious years, long ago and not very successfully.

Formaldehyde

Formaldehyde is an extremely simple chemical compound, which many of us probably remember as the foul-smelling stuff in which dead animals and parts of animals were soaked in high-school biology laboratories. Its chemical composition differs from that of ordinary water by the addition of a single carbon atom in the middle of the molecule; that is enough to make its properties entirely different. It is by far the most important member of a class of chemical compounds called aldehydes, which are produced in the United States by the billions of pounds per year—formaldehyde alone accounts for about three billion pounds. The rest of the world produces even more. Formaldehyde is used in plywood, particle board, foam insulation, other resins and plastics, disinfectants, and of course embalming fluids. (One use of acetaldehyde, one of formaldehyde's siblings, is in the production of acetic anhydride, which is in turn used to produce heroin from

morphine base. The chemical world is closely interlinked.)
The aldehydes are ubiquitous in commerce, so interest in
their health effects has a long history.

Formaldehyde smells awful, and in sufficient concentration (about a part per million) irritates the eyes and
respiratory tract. There is no good evidence that it causes
cancer in humans. Yet it is formally classified as a "probable human carcinogen," and efforts are under way at this
writing to reduce the permissible exposure level in the
workplace by a factor of two or three from its present value
of three parts per million. Why?

An estimated 95 percent of people are sensitive to
formaldehyde at concentrations between 0.1 and 3 parts
per million. The current standard therefore accepts a certain amount of eye and respiratory irritation, and treats
that as harmless. Still, the Consumer Products Safety
Commission tried, in 1982, to ban the use of urea-form-
aldehyde foam insulation in schools and homes, only to
be reversed by a federal court. (Any reader who notices
that there are too many federal agencies in the act will be
right. That's another universal problem—turf wars among
federal agencies.)

In 1984, the EPA invoked the provisions of the Toxic
Substances Control Act (TSCA, passed in 1976), which
empowers it to ban substances, not just carcinogens, that
pose a "substantial risk" of being injurious to the health.
(In one notable exchange in a scientific journal, some critics interpreted the words "substantial risk" to mean "only
the possibility of a probable occurrence," a meaningless
turn of phrase.) The wording of the law distinguishes between natural substances and additives, but that isn't important here.

By now this had become a bandwagon. Shortly after

EPA announced its intention, the Department of Housing and Urban Development (HUD) published a rule requiring that formaldehyde emissions in mobile homes be held under 0.4 parts per million, and, not to be undone, OSHA announced its intent to lower the permissible concentration in the workplace to between 1 and 1.5 parts per million. Finally, in 1987, EPA formally classified formaldehyde as a "probable human carcinogen," based on "sufficient" animal evidence and "limited" human evidence. What is the "sufficient" evidence?

You have seen it. The rodent inhalation test described earlier in the chapter, the most extensive yet made, produced a particular form of cancer (squamous cell carcinoma of the nasal cavity) in rats exposed to the highest concentration, over fourteen parts per million of formaldehyde, six hours a day, five days a week, for two years. But the evidence that these results can be extended to humans comes and goes. At the moment it is inconclusive, and of course humans are not exposed to anything like that concentration, for such extended periods. If they were, they would doubtless exercise an option the rats did not have, and leave the area—exposure to that much formaldehyde is no fun. In fact, it is entirely possible that the rats got their cancers from the constant irritation of their nasal tissues, something that wouldn't happen to people, and not from any specific chemical effect of the formaldehyde.

So the regulatory agencies, uncertain about the facts of the case, are acting conservatively. In the face of clear evidence from rat experiments, they have opted to err on the safe side by limiting the permissible exposure to humans. The fact that formaldehyde has been classified as a "probable" doesn't mean much unless one says just how probable, and not enough is known to do that. The most recent epidemiologic studies show no significant increase in cancers among workers in the affected industries (more than a million workers were studied by the National Cancer Institute), but no one knows what would have happened if the workers had been chained down and forced to breathe fourteen parts per million of formaldehyde, for a lifetime. Nor will anyone ever know.

Further, in this case it isn't even possible to carry out a cost-benefit analysis. While the costs of tighter regulation can be estimated, we don't know if there are any benefits at all. If there are they are not large, since the actual incidence of formaldehyde-induced cancer in people must be far below the threshold for detection. Anyone who says that formaldehyde is known to cause cancer in people is misstating the facts, yet the rodent evidence at large doses is suggestive. Possibly small effects, and certainly no large effects—that will be a familiar story.

Vinyl Chloride

This will be a short story, interesting mainly because it brings out the question of whether the law permits a regulatory agency to even think about the cost to society of the rules it promulgates.

Vinyl chloride, another very common chemical suspected of causing cancer, is classified like formaldehyde, a "probable human carcinogen," but on better evidence. Most familiar in its polymerized form (molecules hung together end to end) as PVC piping, where PVC stands for polyvinyl chloride, it is everywhere. And there is evidence that some people who have lived for a long time around manufacturing plants that emit the vapors have developed a rare liver cancer called angiosarcoma, more often than the rest of the population. In the whole country only a few dozen cases of this disease are seen each year, so they are noticeable. Unfortunately, the same disease is caused by other pollutants, and it has not yet been possible to unscramble the causes; the evidence that vinyl chloride is the culprit is suggestive but not conclusive. The small-animal experiments are more conclusive, but peculiarities in the data leave more uncertainty than usual about the role of vinyl chloride. But those aren't the issues here—this is a legal, not a technical section.

EPA regulates carcinogens (*inter alia*) under both the Toxic Substances Control Act (TSCA) and the Clean Air Act (CAA). Its discretion is different in the two cases, though its job—to protect us from harm—is the same. The laws are very different.

Under the TSCA, EPA must decide whether a substance presents, or may present, an unreasonable risk to human health or to the environment. Note the key word "unreasonable." It must consider the effects of the substance, public exposure to the substance, and the benefits of the substance, so it must carry out a balanced analysis. When considering the issuance of a regulation restricting the substance, it must consider the "reasonably ascertainable economic consequences of the rule," which again means to weigh the pros and cons of the case. Nothing there for a risk analyst to quarrel with.

Under the CAA, the EPA Administrator must keep our air clean. He must regulate "hazardous" pollutants, defined as pollutants that "may reasonably be expected" to cause death or serious illness. How does he go about this? He must establish for each hazardous pollutant a standard "at the level which in his judgment provides an ample margin of safety to protect the public health." One might think that the words "in his judgment" would carry the day in the courts by giving him broad discretion, but that has turned out not to be the case. He is required to provide an ample margin of safety, but nowhere in the law is there any provision that authorizes him to perform a balanced analysis, taking all factors into consideration. Without explicit authorization his hands are tied; he must think only of safety, and nothing else may enter his mind. The legal issue revolves around the meaning of the word "ample."

Once EPA accepted the premise that vinyl chloride was a carcinogen, and that it was impossible to establish any definite threshold of exposure below which it could honestly assert that there was no risk (of course there can never be such a threshold), its goose was cooked. In fact, it overcooked its own goose by openly stating that "it should be assumed, in the absence of strong evidence to the contrary, that there is no atmospheric concentration that poses absolutely no public health risk." Note the absolutely and the strong evidence—the goose is burnt to a crisp. No substance could justify such a strong statement—not even oxygen.

Nonetheless, EPA went on to say that a complete ban on vinyl chloride emissions would close a major industry, and would make no sense, so it reinterpreted the law to mean that one should be reasonable, set the most rigid standard of protection that could be met with modern technology, and let it go at that. Surely, they said, one can only be expected to do one's best. But EPA picked the wrong law, because the Clean Air Act says nothing about the use of reasonable judgment.

It was no surprise that EPA was soon taken to court by the Natural Resources Defense Council (NRDC), which seems to have no qualms about shutting down industries,

and was responsible for the great apple panic of 1989.
NRDC argued that, since the EPA admitted that it could
not certify an emission level below which there was no remaining risk, it must set the emissions limit at zero. That
would have ended the production and use of vinyl chloride
in the United States.

The court began by observing that many other pollutants were without a known damage threshold. They
thought it highly unlikely that Congress meant to shut
down large parts of American industry when it passed the
Clean Air Act, without even noticing what it was doing.
The last part follows from the fact that no one seems to
have mentioned the possibility of major economic disloca-
tion during the congressional debate that led to the passage of the law. The court presumed that Congress was
sensitive to the impact of the laws it was passing—a commendable presumption.

In the end, however, the court found no justification
in the law for the use of anything but health criteria to
set standards. While the judges did not agree with NRDC
that the emission standard should be set at zero, they did
rule that a criterion that cited the "best available technology" was not based on health criteria, and was therefore
inadmissible under the Clean Air Act. So they sent the
Administrator of EPA back to the drawing board, to find
a health-based rationale for doing what he originally intended to do. They did not prohibit him from setting a
reasonable emissions limit, but required him to cite only
reasons within his discretion, in determining an "ample
margin of safety." The forbidden considerations include
cost and technical feasibility. It is just a bit bizarre to
resolve a technical matter without considering cost and
technical feasibility. But NRDC had succeeded—balanced

risk assessment had lost again.

The court's decision in this case is notable for an interesting and quotable footnote:

> If the Administrator finds that some statistical methodology removes sufficiently the scientific uncertainty present in this case, then the Administrator could conceivably find that a certain statistically determined level of emissions will provide an ample margin of safety. If the Administrator uses this methodology, he cannot consider cost and technological feasibility: these factors are no longer relevant because the Administrator has found another method to provide an "ample margin" of safety.

Stripped of its opacity, this says that the Administrator may not, under the law, consider cost or feasibility in setting emission standards. However, he is invited to invent some statistical obfuscation to provide a rationale for his actions, provided it is based only on health considerations, and doesn't use the forbidden c- or f- words. For those with long memories, this opinion was written by Judge Robert Bork, of the United States Court of Appeals for the District of Columbia Circuit.

Asbestos

Strictly speaking, asbestos may not belong in a chapter on chemical carcinogenesis; the mechanism through which it induces cancer is not yet understood, and may be physical or chemical. It is surely an irritant to lung tissue, which is serious enough, and exposure can lead, in time, to a malignancy. Whether physical or chemical in origin, the cancer isn't any less real. Asbestos does more harm to smokers than non-smokers, presumably because the lung damage is cumulative, but there may be deeper reasons for the synergism. Although the carcinogenic characteristics of asbestos will be emphasized in this section, it can lead to a number of other diseases; though not cancer, they also kill. And there are different varieties of asbestos, with different effects.

Much of the exposure to asbestos has been occupational, and nearly all of that in the past. Because asbestos is fireproof and durable, it was widely used for such items as theater curtains, car clutch and brake linings, building insulation, and roof shingles. It was also common in such industries as shipbuilding, especially during World War II. Whatever the need for a fire-retardant, abrasion-resistant, mold-free, and tough substance, asbestos was the material of choice. It is a fiber woven of long thin rock crystals.

The story of asbestos extends far back in human history, simply because it is so durable. It is a mineral, and surface deposits of the rocks from which it comes were within easy reach of primitive man. According to the National Academy of Sciences, it was used for Egyptian embalming cloths, for Roman cremation wrappings, and by the Romans for everlasting wicks in the lamps of the Vestal Virgins. Charlemagne, over a thousand years ago, is supposed to have had a tablecloth made of asbestos; he could clean it after a feast by throwing it into the fire.

Now that we are more aware of the damage asbestos can do to human health, it has been substantially removed from commerce. The residual risk is in the buildings that still contain it, and in the people who bear the legacy of damage. Theater curtains are no longer made of asbestos, new home and school insulation no longer contains it, and other materials are used where fireproofing is necessary. So the risk is fading away in the United States. World production, on the other hand, has held steady or increased over the last few decades, our share decreasing to about 1 percent. The few remaining uses are being phased out in the United States.

Despite its decline here, asbestos is interesting for two reasons. The first is that our country is in a barely controlled frenzy of rooting the asbestos out of public buildings, mostly schools, for fear that it may be disturbed at some time in the future. Of course the very act of rooting it out is bound to release fibers that might otherwise stay in quiet repose for eons.

The other point of interest is that the widespread realization in the 1960s and early 1970s that asbestos could be harmful has loosed a historic tidal wave of personal injury claims. More than thirty thousand lawsuits have been filed in the nation's courts, with claims totaling many billions of dollars, focusing attention once again on the issue of product liability. Should a company or an individual be held financially and legally responsible for producing a material that was generally considered safe at the time, but was later found to be otherwise? If history judges us in retrospect to have made an honest mistake, are we guilty and punishable? Put in those terms, most people would probably say no, but the answer in the courts is more frequently yes. Juries seem to be quite willing to take money from companies and governments, and give it to individuals. Since about half the exposure in asbestos litigation is associated with work in government shipyards, this opens up the deepest of deep pockets. Of the loot from asbestos lawsuits, the attorneys for both sides receive about 61 percent, and the victims about 39 percent.

We needn't belabor the point, but it *is* important. As knowledge and understanding of risk improve, not only does our world become safer, but also more vulnerable to manipulation. Peter Huber, a lawyer and engineer, has written: "Give me a scientist who is willing to put a number—any number—on the risk of dying, and I will give you a plaintiff's attorney ready and willing to take the number to court." The more information available on a risk, the better it can be slanted in litigation, especially if honest scientists admit uncertainty, and it is then almost a truism that honesty is not the best policy.

Asbestos is not unique in these matters. There has been considerable pressure in the last decade or so, fiercely resisted by the various trial lawyers' associations, to reform the nation's tort laws. (In legal jargon, a tort is a civil wrong inflicted on one party by another. It can lead to a liability for damages.) The combination of an increasing propensity to sue, the effect of the "deep pockets" syndrome in increasing liability awards, and a greater willingness to use product liability as a form of income redistribution, has had a major effect on the economy. There are many examples. Medical malpractice insurance is a familiar case, having forced some obstetricians to abandon their specialty, and raising medical costs for everybody. (Many medical tests are now made to protect the doctor from legal problems later, not to benefit the patient.) Over three hundred thousand damage claims were filed in the famous Dalkon Shield case, a record. One-tenth as many general aviation aircraft are now produced as were manufactured only ten years ago, and the average liability insurance burden on each one is a hundred thousand dollars. And so on.

For asbestos, the knowledge that there was risk has not changed much in the last two decades, but awareness that there were ways to be compensated has. It was not because of sudden discovery of the carcinogenic potential—that gradually became clear to everyone between about 1955 and 1975, and indeed some life insurance companies stopped selling policies to asbestos workers as early as 1918. That there is real risk in exposure to asbestos fibers has long been known. It has been estimated that about five million people had some exposure through work in the World War II shipyards. We don't know with any certainty how many were damaged.

As a general rule, the airborne asbestos count is lower in rural areas than in cities, and lower in cities than in factories that deal with asbestos. It is of course higher in buildings that have been built using asbestos-containing materials than in those that have not. And the risk increases with the length of the exposure. Cancer from exposure to asbestos typically takes over ten years to develop.

Two distinct classes of people merit concern. There are the exposed workers who bear the marks of long exposure to high concentrations, at a time when the country wasn't so careful about workplace risk, i.e., before OSHA. Then there are the schoolchildren, for three reasons: we care about our children, and that hardly needs emphasis or apology; they spend a good share of their lives in their schools; and they are young, so any damage has longer to develop into a real problem.

Happily, average school exposures are far lower than had been thought when this all began. (EPA itself reduced its estimate by a factor of ten between 1983 and 1985.) In fact, the exposures in school buildings average less than a thousandth of the limit that OSHA has placed on workplace exposure. For the vast majority of schools, the expert consensus is that the best thing that can be done for the children is to leave the stuff alone unless it is exposed and friable, not to release it to the atmosphere in a clumsy effort to get rid of it. That is not a popular observation.

For the exposed workers, it is different. Many have suffered substantial damage; little can be done to improve their health, though something can be done to compensate them. But at whose expense? And how does one separate the needy from the greedy? The dilemma is familiar in our increasingly litigious country.

What has litigation to do with technological risk? It is relevant because the potential for gain by lawsuit provides social and economic motivation toward the exaggeration of risk, particularly toward the exaggeration of technological risk, where the picking is easier. That is one factor forcing the regulatory agencies to err on the extremely conservative side when assessing risk, which, in turn, increases the risk of life for all of us. The reason many schools and businesses are removing old asbestos is fear of later litigation, not fear of asbestos. Asbestos in good condition is best left alone, both for health and cost reasons. It has been estimated that we are spending over a billion dollars to save each life by removing asbestos from buildings, and a billion dollars can save lives more effectively in other ways. It is sometimes better to let sleeping dogs lie.

13
Highway Safety

FAMILY CHARACTERISTICS

A hundred thousand people die accidental deaths each year in the United States, almost half of those in road accidents. In certain age groups—males in their teens and low twenties—automobile accidents are the leading cause of death. And there are five million injuries per year, costing about $50 billion in medical and non-medical care for the injured. That the roads are risky needs no proof.

Still, some perspective doesn't hurt. The combined death rate from murder and suicide is about the same as the highway death rate, and is increasing while the latter is decreasing. Unlike highway transportation, little can be said for the social benefits of murder and suicide. Smoking kills nearly ten times as many as road accidents, and not much more can be said for its social benefit. Traffic accidents provide about 2 percent of our total mortality. In fact, the United States has the lowest fatality rate, per passenger-mile, of the major industrialized countries, but we also drive more. So road travel carries risk, but is not the ultimate threat to human survival.

And it really has been getting safer on the roads. Though the total number of fatalities hasn't decreased much in the last twenty years (it is about the same now as in the mid-1960s, when Congress noticed the subject), the fatality rate *per passenger-mile* has gone down by more than half. It is widely believed among experts that the trend among the states (now complete) to raise the minimum drinking age to twenty-one had a great deal to do with that. Probably the trend toward wider use of seat belts helped, as did the speed limit, though the decreasing trend predated both of these. As a national average, an occupant has about one chance in a hundred million passenger-miles of being killed in an automobile accident. For a person on the road for ten thousand miles a year, that adds up to about one chance in ten thousand per year of being killed in an automobile accident, much larger than many of the other risks we try so hard to exorcise, but still less than a tenth the average risk of death from smoking.

As with all risks, some population groups are more affected than others. The death rate for white males in the 15–24 age bracket is a full three times the national average, while for black females in the same age bracket it is half the national average. Of course these extremes reflect cultural and social differences, exposure on the road, and other imponderables. They also reflect the effects of youthful drinking.

The drinking age has long been an issue muddled by extraneous factors. For many years matters like the requirements (including age) for voting and for purchase of alcoholic beverages were regarded as within the province of the individual states, and federal interference was min-

imal. During World War II, however, and until the end
of the military draft in 1973 after the agony of Vietnam,
eighteen-year-olds were eligible for draft into the armed
services. This led to the feeling that there was something
incongruous about a situation in which a person thought
to be old enough to fight to defend the country was still
too young to participate in its government. Of course, the
skills required to do these two quite different things are entirely distinct, but there was still a sense of injustice—an
itch that needed to be scratched. So the XXVIth Amendment to the Constitution was ratified in 1971, lowering the
minimum voting age to eighteen, in all the states. Inter-
estingly, the Amendment doesn't say quite that. It is still
permissible for a state to set a still *lower* voting age, like
five, but not a higher one.

A number of states then lowered their drinking age (as
it is called, though it refers to the minimum age for the
purchase of alcoholic beverages), arguing that a person old
enough to vote ought to be old enough to drink. (Since the
argument that he was old enough to vote was that he was
old enough to fight, the syllogism implies that if he is old
enough to fight he is old enough to drink. Not many would
carry the logic that far, but logic it is.) None of this discussion, of course, has anything to do with highway safety.
As late as 1984, when the federal government finally took
action, the minimum drinking age varied from seventeen in
Arizona to twenty-one in exactly half the states. The others were scattered between these extremes, with eighteen
a popular drinking age. In 1984 a federal law was passed
that did not require that the states raise their drinking
ages—that might not be constitutional—but penalized the
recalcitrant through the loss of federal highway funds.

One by one the states toppled, Wyoming capitulating

in 1988, so there is now a national standard drinking age of twenty-one, enforced by tax money. This will doubtless have a beneficial effect on the highway fatality rate, especially for younger drivers, since that has generally been the case in those states that made the move earlier. Wyoming held out longest for a variety of reasons, even under the threat of loss of federal highway funds, but was loath to be the only state in the Union with a lower drinking age. It had been nineteen.

Drinking drivers don't carry the only contribution of alcohol to the highway death toll. To the extent that pedestrians are killed too (almost 20 percent of fatalities are pedestrians) their drinking contributes. A full one-third of pedestrian victims have blood alcohol concentrations above the legal definition of drunkenness.

But of course alcohol isn't the only, or even the dominant, source of highway risk, although it is a real contributor and satisfies a natural craving for easy solutions to hard problems. Especially for people who believe that alcohol is evil anyway. Let's look at the facts.

First, though the annual carnage on the highways led in 1987 to about 46,000 fatalities, fewer than three-fourths were vehicle occupants. The remainder, no less dead, were pedestrians, motorcyclists, and a few smaller categories. (The number of motorcyclist fatalities has bucked the trend and gone up dramatically in the last twenty years, perhaps because of the proliferation of affordable imported motorcycles.) Of the vehicle occupants, about two-thirds were in passenger cars, the remainder in buses, trucks, and pickups, mostly the latter. Thus, occupants of passenger cars constitute just about half of all highway fatalities. About 27,000 actual drivers (17,000 driving passenger cars) died in 1987, of whom about 7,350, 27 percent, were between eighteen and twenty years old. The fraction of the population in that age group is less than 5 percent, so they contributed more than five times their proportionate share to the death rate.

The point of this statistical recitation is that the highway safety problems are not as simple as some would have us believe. They are influenced by age, alcohol, vehicle, road, and a host of less tangible factors, including social attitudes toward driving. As for alcohol, nearly half of all licensed drivers involved in fatal accidents had been drinking, and more than one-third (38 percent) of those killed were over the legal alcohol limit. The peak for this misbehavior is in the 20–24 age bracket, where nearly half of all fatal accidents involve a driver who has been drinking, but that same percentage hovers around 40 percent up to about age fifty-five, at which point it begins to decrease. Even if the role of alcohol is exaggerated by the temperance folks, there is no question that it is a real problem on the highways, especially among the young. By the time we get to the sixty-five-and-over group, only 10 percent of fatal accidents involve a drinking driver. Either old folks don't drink and drive, or those who don't drink and drive grow old. Reader's choice. The percentage of departed motorcyclists who have been drinking is higher still.

Most states that have raised the drinking age in recent years have experienced a small but real reduction in traffic fatalities. It is rarely as big an effect as the enthusiasts would have us believe, but it is there.

That leads to the central question of whether increased regulation would help, whether through punitive measures against speeding or drinking drivers, or tighter standards for DUI (driving under the influence), or lower speed limits (they were as low as 35 mph during World War II).

Each would doubtless improve safety, but in each case there would be a cost to society in return for the increased safety, and those costs would have to be compared to the benefits.

In this introductory section there has been no mention of the other factors that contribute to risk, like crashworthiness of vehicles, guard rails, lighting and posting of highways, controlled access, and such matters. Some of the principal points are illustrated by two contentious issues, mandatory seat belts and the 55 mph speed limit. The former has to do with crashworthiness, the latter with driver habits.

Seat Belts

It is axiomatic in the traffic safety business that every automobile collision is really two collisions. First the car collides with some other object—another car, a guard rail, a lamppost, or the ground at the bottom of a cliff. There follows the collision of the occupants with either the interior of the car or the ground or something equally inhospitable. Restraints are directed at mitigating the second collision by keeping the passenger in the car, while substituting something softer to hit, like a seat belt or shoulder harness, or an air bag. The concept has been around for a long time—the first patent for a seat belt was issued in 1885.

No one doubts that seat belts would work if used, if by "work" we mean reduce fatalities in collisions, but they don't seem to have the effectiveness in practice that is claimed in theory. Tests on dummies in simulated crashes, as well as observation of real accident victims on arrival at the hospital, both suggest that 40 to 50 percent of unbelted people killed in collisions could have survived if they had been belted. Yet the improvement in states that have adopted mandatory seat-belt laws has been closer to 5 to 10 percent, if that much. Why?

In general, improvements are rarely as great as the promoters promise; it is hard to sell anything nowadays, even a good idea, unless the value of the product is overstated. Then, when it is sold, and isn't the panacea that was originally claimed, that will have to be explained away. This applies to safety measures and presidential elections, and is a consequence of the long reach of Gresham's Law.

That is worth a paragraph, even though it has little to do with seat belts. Gresham's Law, an economic rule named in the eighteenth century after a prominent sixteenth century English merchant (under the mistaken impression that he first stated it), asserts that if two kinds of money are available, the bad money drives the good money out of circulation. If there is gold around, and the government issues "worthless" paper money, people will buy their bread with the paper and will bury the gold. The least valuable medium always wins the head-to-head contest. The depressing applicability to the wider world of ideas is apparent everywhere, to those who watch commercial television, to those who browse among the magazines at newsstands, to those who listen to political debates, etc. For risk matters the debate is more often than not dominated by those who have easily understood solutions, easy-to-hate villains, and painless cures. In other words, counterfeit money. Recall the Mencken quotation in Chapter 4.

There are two simple theories, each with at least a few adherents in the traffic safety community, for the disappointing effectiveness of seat belts. The first, with the elegant name of risk homeostasis, is not very popular among the experts. It is also called risk compensation. Home-ostasis in the animal and plant worlds is a tendency for an organism to respond to changes in its environment by adjusting to maintain the *status quo ante* as best it can. In the world of risk this would mean that, deep in our blackest hearts, we somehow know how much risk we want to live with. Therefore, if the outside world, i.e., the government, imposes safety devices on us, we react by acting more carelessly, in order to maintain our accustomed level of risk. It's an engaging thought, clearly with at least a germ of truth in it.

According to this theory, a driver who buckles up (and whose passengers, if any, buckle up—most cars have only the driver) will drive a bit faster, take a few more chances, and generally be more prone to complacency than he would otherwise be. And complacency is the archenemy of safety. Some limited but ambiguous evidence suggests that that is the case. It would be nice to observe the same driver both before and after he started using his seat belt, just to see if he really takes more chances, but that is very difficult. It has not been done.

The majority of risk experts are not taken with the idea of risk homeostasis; indeed they dislike the whole idea. One does, however, have to be careful in interpreting that fact. If there really were an important tendency in people to compensate for well-intended safety measures, it would render much of the work of the safety community point-less. That would be unthinkable, certainly to the safety community. Risk homeostasis *is* a real effect—high-wire artists take more chances with a net under them—but per-haps not relevant to this question. One can only say that more research is needed. Those words are music to the ears of a scientist.

The second theory for the limited effectiveness of mandatory seat-belt laws has again a solid base in human behavior. Unthinkable though it may be, people have been known to disobey the law. Before the wave of mandatory seat-belt laws in the mid-1980s, estimates of the fraction of drivers using their belts ranged around 15 percent— the belts have never been very popular. Most states that passed the laws requiring people to buckle up noticed an initial surge of compliance, up to roughly 50 percent—the states vary—which then dropped off after a few months. The remaining argument for air bags or other really passive, i.e., compulsory, restraints is based on those numbers. By golly, if people won't voluntarily do as I say, I'll make them do it. For their own good, of course.

There is also a difference among the states in the vigor of law enforcement. So-called primary-enforcement states permit a policeman to arrest a motorist going by without a fastened seat belt, and compliance is understandably higher in such states. In secondary-enforcement states, a driver must be stopped for another alleged infraction before his belt usage is subject to check. It is add-on enforcement in such states, and compliance is lower. Most states are in the second category; some have even repealed their seat-belt laws, under popular pressure. Seat belts may be life-saving, and are passively accepted by many, but they have few real friends.

The second ineffectiveness theory suggests that the drivers who don't wear seat belts are most likely to have accidents. In this picture, fatal accidents are concentrated in a more antisocial subset of the population: less likely to be reached by educational campaigns, more likely to drink, more likely to drive carelessly, and more likely to ignore the seat-belt laws. The argument is that the careful and responsible drivers were wearing seat belts before it became mandatory, the law simply causing the next most careful drivers to buckle up. Under this theory there would be no effect on the habits of the really accident-prone. The concept is called selective recruitment.

Safety statistics lend some support to this idea. Of the passenger car drivers who were killed in accidents in 1987, 8 percent of the drunken drivers were wearing seat belts, compared to 24 percent of the sober drivers. (Statistics must always be treated with care. These data could be interpreted to mean that drunks not wearing seat belts are better able to avoid fatal accidents. But that would be wrong, wouldn't it?) A number of studies have shown that accident-prone drivers really do have less of a tendency to buckle up, so there is little doubt that some selective recruitment is present, but more research is needed to determine how important it is. More music.

Seat belts are a mixed bag—no pun intended. The theory is solid, and the belts would save many lives if universally used, but the actual results in practice don't validate the theory. And people don't like them. It is easy to say that people who don't like them are irrational, but they are in effect asserting their own values over those of their protectors. That may offend the well-meaning protectors, but it is not irrational. The current plan is to require tamperproof restraints in all cars, probably in the form of passive seat belts. These wrap themselves around the motorist, willy-nilly, when he closes the door and starts the engine. Air bags, the leading alternative, have the advantage of being unobtrusive until they go bang, and the disadvantage of being more expensive. Whether the driving population will accept passive seat belts with more docility than it did the interlocks in 1974 is anyone's guess.

55 mph

If those who were under ten years old at the time remember very little of the episode, then 40 percent of our population has no memory of the Arab oil embargo of 1973–74, and the trauma that accompanied it. The United States was importing about half its oil, mainly from the Arab countries in the Middle East. This oil supplied gasoline for our cars, fuel for our airlines, heating oil for our homes, and so on, and the five-month boycott was a major blow to our lifestyles. We really hadn't appreciated our dependence on imported oil. Every effort was then made to conserve supplies (even though the boycott leaked—some countries can't keep promises to each other), and one such measure was the 55 mph speed limit throughout the country. It was imposed as a matter of national emergency, and became fully effective at about the time the embargo was lifted in 1974. Even so, it has persisted to the present, mainly because regulations have a life of their own, and develop their own *raison d'être.* In this case the rationale was safety, though safety was hardly mentioned when the limit was first imposed.

As it was for seat belts, no one doubts that lower speeds can save lives, and again it is hard to determine just how well it has worked in practice. Safety as the reason for the speed limit is an afterthought—had it been suggested originally for that purpose it would surely have been rejected. Judging from the average speed at which motorists travel in 1989, even on roads that are posted for 55 mph, the speed limit is not now taken very seriously by many drivers. There seems to have been a collective judgment of the American people that risk is preferable to delay, if we measure their judgment by behavior, not by words. It is bad practice to have laws that aren't enforced—it breeds contempt for the law and gives too much judgmental authority to the enforcer of the law. It therefore provides the opportunity for discriminatory enforcement, whether against small red cars or large green people. That is bad for democracy.

The reason that lower speeds are expected to reduce fatalities in collisions is solidly grounded in physics. The damage done in a collision—whether with another car, a signpost, or a bullet—is roughly proportional to the square of the impact velocity. Double the impact speed and the damage is quadrupled. A reduction from 70 mph (the previous limit on Interstate Highways) to 55 mph can reduce the damage potential of a collision by 40 percent. That seems worthwhile. Of course the high-speed Interstate Highways have about half the fatality rate of slower roads, so speed isn't everything.

As mentioned earlier, the highway fatality rate per passenger-mile has decreased substantially in the last fifteen years. There have been some ups and downs, but down has been the norm.

Much has happened in those years: the Arab embargo, which greatly but temporarily reduced the amount of discretionary automobile travel; seat belts; the increased price of gasoline; improved crashworthiness of cars; growing safety awareness in the country, which spilled over into driving; increased drinking ages; and continued development of the safer Interstate Highway system. Each of these can be given some credit for the reduced fatality rate, but it is hard to honestly apportion the credit among them. All have helped.

In particular, we don't know how much benefit to attribute to the speed limit. Whatever its contribution, it comes at a cost which, again judging from their behavior on the roads, people don't want to pay. In that sense, the speed limit on the major freeways and the Interstate Highways shares some features with the Prohibition era of 1919–33. People can be coerced into conformity, but if the punishment is bearable the social norm is to seize all opportunities for non-compliance. Of course, if the punishment is then made unbearable, people will obey—that is the experience of countries with other forms of government.

Strangely, a substantial majority of respondents to polls say they support the 55 mph speed limit. At the same time, they exceed it as a matter of practice, exercising their right to say what is expected rather than what they believe. (That is an adaptive skill learned early in life.) Estimates suggest that drivers spend approximately a billion extra hours on the road, per year, because of the speed limit, little of it enjoyable. There are also regional differences; many Western drivers know what it is to spend long hours on straight roads, with visibility as far as the eye can see, wondering why they are expected to crawl. Since patrol cars are also visible on such roads as far as the eye can see, there is little adherence to the 55 mph speed limit. The *average* speed on Interstate Highways posted for 55 has been measured in a few places, and is about 60. That suggests that quite a few go faster, consistent with casual observations. However, high technology may yet win the day through aerial surveillance of speeders.

Recognizing the inevitable, Congress passed a bill in 1987 permitting the states to post up to 65 mph speed limits on certain rural Interstate roads, and within six months three-fourths of the states had done just that. The bill was only barely passed, yet it was responsive to public distaste for the limits, and the Congress is exquisitely sensitive to public tastes and predilections. Still, only a small fraction of roads have been affected, and the eventual consequences of the higher speed limits are uncertain at this writing. The first reports showed mixed results, varying from state to state. California is a good illustration, since it posted 77 percent of its eligible Interstate miles at 65 mph, leaving the rest at 55. Just looking at the rural Interstates, the fatalities on the sections for which the speed limit was raised went up from 110 to 116 between 1986 and 1987. On the sections for which the speed limit was kept at 55, fatalities went from fifty-six to sixty-five in the same period. The difference isn't statistically significant (doesn't pass the square-root-of-N test), and is certainly not dramatic.

More recent data come from New Mexico, whose rural Interstates showed an average of sixty-one deaths per year for the five years before the change to 65, and ninety-nine in the subsequent year. That is statistically significant. Before the change, about 15 percent of cars were going over 65 on these roads (while the speed limit was 55); that went up to about 30 percent, though the median speed only went up about 3 mph. New Mexico is a special case, a thinly populated state with fast roads. More data, from many states, will be coming in over the next few years.

What are the benefits and costs? The numbers that follow are drawn from a reading of several studies, and are simply guesses by the author of which is the most credible estimate for any variable. As with seat belts, the problem is that the many changes since 1974 make it impossible to separate out the effect of any given change, including the speed limit. In addition, opinions in this business are strongly held, and can influence the conclusions drawn from solid data. Some readers will have noticed that many

European countries think we overemphasize the speed issue to the point of fanaticism; many of their major highways have no speed limits at all. Driving on those roads
can be pretty exciting.

To begin, it is useful to understand the role of the
high-speed roads in the overall transportation picture. Although the Interstates and freeways posted (before 1987)
at 55 mph constituted less than 1 percent of our four million miles of paved streets and highways, they carried a
fourth of all the traffic. Such roads have always been safer
than other roads—turnpikes and Interstate Highways have
a fatality rate much lower than the national average. According to the National Highway Traffic Safety Administration (NHTSA), the 1986 fatality rate on all U.S. highways was 2.5 fatalities per hundred million vehicle-miles;
the comparable number on the rural Interstates was 1.1,
less than half as high.

The average speed on the Interstates went down in
1974 when the limits were imposed, but has been gradually creeping up ever since. However, the fraction of drivers
traveling at extremely high speeds, far above the speed
limit, decreased sharply, and has not yet returned to the
pre-1974 level. The nation's preoccupation with excessive
speed seems to have gotten around. Thus accidents involving those really speedy cars are likely to be less lethal
now than they were. However, cars are more crashworthy
than they were; the minimum drinking age has gone up;
the population has aged (the fraction of the population between the ages of eighteen and twenty-one decreased from
7.7 to 6.3 percent from 1975 to 1986, nearly a 20 percent
decrease), and youth contributes more than its fair share
to the accident rate. Any one factor can be made to look
more effective by ignoring the other contributors.

Also, increasing the speed limit doesn't necessarily translate into increasing speed. Many drivers seem to set their own speeds, and comply with the speed limit only as necessary to minimize traffic arrests. For such drivers the effect of an increased speed limit is to legitimize the speed at which they have always been driving. Thus, an increase of 10 mph in the speed limit on the rural Interstates, from 55 to 65, actually resulted in an increase of average speed of less than 2 mph, from 59.7 to 61.4, magically turning a majority of drivers from lawbreakers into law-abiding citizens. It is easier to change the law than to change people's behavior. For the 85th percentile (the speed exceeded by only 15 percent of the drivers) the increase was also less than 2 mph, from 65.6 to 67.3.

With those caveats in mind, the best estimates seem to be that the reduced speed limits were responsible for saving about two thousand lives per year, give or take a factor of two. The other benefits (other than reduced injuries) are truly negligible in the grand scheme of things, consisting of fuel and other economic savings, and amounting to at most a few hundred million dollars per year. That may sound like a lot, but it is only a dollar or two per car.

Similarly, the cash costs of the 55 mph speed limit are not high, again amounting to some hundreds of millions of dollars per year, much of it in the form of reduced efficiency in the interstate trucking business. Of course, interstate truckers, armed with CB radios to frustrate law enforcement, are notorious evaders of the speed limits. They are also skilled drivers.

So the substantial benefit of a speed limit is the saving in lives, and the substantial cost is the value of time lost. Experts on both sides of the fence will call this a great oversimplification, and so it is, but complexity sometimes obscures a point. Not many cost-benefit questions are so simple.

Using the slightly-better-than-a-guess estimate of two thousand lives saved, and the somewhat better estimate of a billion hours lost per year, it is easy to make the comparison. The annual income in the United States in 1987 was approximately $18,000 per capita, including everything, and if the value of people's time is estimated by dividing that by the number of hours in a working year, it turns out to be about $9 per hour. Of course there should be adjustments for small children and for the fact that personal time is never as valued as working time (who has never estimated that his time is so valuable that he can't afford to mow the lawn or paint the kitchen himself, but then done it anyway?), but all such adjustments, some up and some down, are niceties. An adequate compromise is to take something like $10 per hour as a reasonable estimate of the value of time for drivers and passengers on the relevant highways, with due respect for others who, with more elaborate methodology, have used numbers ranging from the minimum wage up.

If we then divide $10 billion for the value of time lost (a billion hours at $10 per hour) by two thousand lives saved, we find that each life saved is costing approximately $5 million. That is substantially more than is spent in most other regulatory activities. The argument that this kind of life saving is free of cost consists plainly and simply of ignoring the value of other people's time. Perhaps, in some obscure way, the American people understand this, and that is why they have tended to ignore the law.

14

Air Transportation

FAMILY CHARACTERISTICS

It is truly a miracle on which few of us ever ruminate that we are able to fly through the sky at speeds a hundred times faster than we could have traveled a little over a hundred years ago, in reasonable comfort and safety, watched over by solicitous cabin attendants, and more often than not arrive at the intended destination on the appointed day. In the first reported air travel, a couple of thousand years ago, there were no creature comforts, and the fatality rate was 50 percent. According to the legend, Dædalus survived, but his impetuous son Icarus did not. We have come a long way.

For orientation, consider this wonderful article from the *American Review of Reviews* of November 1909, which sings the praises of the recent accomplishments in aviation, and then goes on to say:

> So much for the achievements of the last two years.
> It is truly a marvelous record, far exceeding the

expectations even of the enthusiasts. And yet it must be evident to any thoughtful and impartial observer that, great as these performances are, they do not by any means justify the extravagant claims which have been made for this new and fascinating toy,—for toy it is, at least in its present state of development. There is a long road to travel before it becomes available for the average sportsman and takes its place with the automobile as an established means of recreation.

The article continues by debunking all the crazy claims that have been made for aviation, especially the claim that it might be useful in warfare. So much for foresight.

The current fatality rate for passengers in commercial jet aircraft is approximately one fatality per *billion* passenger-miles, ten times better than the average for automobile driving, and about a hundred times better than the average for general aviation. The safety record of the large commercial airlines, measured by a passenger's chance of getting killed when he or she steps into one of their airplanes, has improved by a factor of two hundred in the last fifty years, and there are parallel improvements in the non-airline aviation world. In the same period, the fatality rate for motor vehicle travel, per vehicle-mile, has decreased by less than a factor of five. (This author ought to make one admission at the outset. As a general aviation pilot for over forty years, he has a warm feeling toward aviation, and this may conceivably color his views. He will try diligently to keep this from affecting his objectivity, but it is for the reader to judge how well he has done.) Anyway, commercial aviation is truly the safest known way to travel long distances, bar none.

Even though these statistics are correct, the fatality rate in commercial aviation varies wildly from year to year, depending on whether or not there has been an accident. There were no fatal accidents in scheduled airline service in 1980, while 1979 saw the worst domestic aircraft accident so far (the DC-10 at Chicago, in which 275 died). There were 233 fatalities in 1982, while 1984 produced 4. The 197 fatalities in 1985 were followed by 3 in 1986. And all this with over seven million flights per year, carrying more than four hundred million passengers an average of a thousand miles each. Though it is easy to average these fatality numbers, and that is what we have done, any given year may have a major accident or none at all, or even several—that is the nature of statistics. When there is an accident there are cries for improved safety, when there is not we take it all for granted and complain about airline food.

Commercial aircraft accidents involve such large numbers of fatalities when they do occur that they attract public attention, and people are misled into believing they are more frequent than they actually are. Averages involving rare events are hard to visualize, but we still have to be quantitative about risk. A major report to the President in 1988 opened by saying that the "Commission unanimously concludes that the nation's air transportation system is safe," thereby falling instantly into the trap of believing that safety is like a kitchen spigot, on or off. It isn't, and like many real kitchen spigots it leaks; what matters is how much it leaks. Lord Kelvin once said, "When you can measure what you are speaking about, and express it in numbers, you know something about it." Those who say something is safe or unsafe are avoiding work. That includes presidential commissions.

Though the accident rate is much higher in general aviation (with charter flights and commuter airlines not quite
as high), all flying has been getting safer since the early
days of aviation. Contributions to improved safety have
come from all parts of the equation: from the design, construction, and maintenance of aircraft; from the improved
reliability of engines; from the dramatic improvements in
electronics that have made operation in bad weather so
easy that even mediocre pilots survive (thereby frustrating
the laws of natural selection, at least among pilots); from
higher operating altitudes; from improvements in weather
forecasting; from a continuous learning process about the
sources of risk; and from a host of other contributors. Only
the air traffic control system has resisted change in the last
few decades, using new technology only to do what it has
always done, more efficiently and to more aircraft. That
is why the system always functions as if it were on its last
legs.

The air traffic control system has one basic objective:
to keep aircraft from colliding with each other, *when they
are under control.* It has no responsibility for keeping
them from hitting mountains or the ground, or running out
of fuel, or unintentionally flying upside down—those are
the responsibilities of the pilot. The system is owned and
operated by the Federal Aviation Administration (FAA),
which (uniquely among regulatory agencies) has responsibility for both regulating *and* promoting aviation, as well
as operating the very system it regulates. It does this with
its own employees, the controllers. The FAA is now part
of the Department of Transportation, but there are moves
afoot—the Aviation Safety Commission mentioned above
recommended it—to restore its earlier independent status.
The current arrangement is both historic and bizarre.

We should look at the numbers. The FAA has about 50,000 full-time employees, and consumes an annual budget of about $5 billion. It controls the lives of a community of about 700,000 licensed pilots, of which about 150,000 have commercial licenses, 80,000 with airline transport ratings. There are about 200,000 active aircraft of all types in the country, of which about 3,000 are jet aircraft operated by the airlines. The scheduled airlines carry three times as many passengers as general aviation, in a tenth the number of flights, and carry them further and faster, more safely, but to fewer destinations. There are over 10,000 airports in the country, of which only about 400 are served by scheduled airlines, most of the remainder by general aviation. It is a rich and complex mix of equipment and people.

The FAA operation has always been labor-intensive, the controller on the ground having responsibility for an aircraft in flight (provided that the flight is under any control at all—most general aviation flights, which is to say most flights, aren't), and passing the responsibility to the next controller in line as the flight progresses. There is a polite fiction that the instructions a controller gives to a pilot in flight are only advisory, that the pilot is really responsible and in command. This is meant to help the FAA to evade legal responsibility when the controller makes a mistake. Any pilot quickly learns the real facts of his relationship to the controller, which are that the controller is running the show. In fairness, the controllers generally do their jobs courteously and well, and are a devoted lot.

The fundamental rule of flying has always been see-and-be-seen, which is applicable when the weather is good and it is possible to fly under visual flight rules (VFR). When the weather is not good, instrument flight rules (IFR) apply. It is only under those conditions, or in special places like high altitudes or near large airports, that the FAA becomes an active participant in the drama of flight. Its historic job is to keep aircraft from colliding in the air when visibility is bad and see-and-be-seen is inadequate. In fact, the airlines always fly with IFR flight plans, whatever the weather or visibility, so they are always under what is called positive control. The airlines first adopted this practice long ago, in a move to overload the system and thereby force the FAA to hire more controllers and provide more service. It worked. Of course, even when flying with an IFR flight plan, a pilot is required to maintain a visual lookout if the visibility permits. Some pilots are more diligent than others.

By 1980 there were 27,000 controllers working for the FAA, to service the average of 700 airline aircraft and four times as many general aviation aircraft in the sky, over the entire United States. These numbers vary both in space and time, and most general aviation aircraft are normally not flying under instrument rules. Most general aviation pilots are not even qualified to do so. Even with five shifts, therefore, the ratio is several on-duty controllers for each airplane in flight. There are other duties beside flight-following, but this is still one measure of a labor-intensive organization.

In 1981 the controllers' union called an illegal strike, the President condemned the strike and fired the striking controllers, and their numbers were reduced to 17,000. That was a golden opportunity to move quickly with plans to modernize the system, and to improve the ratio of technology to personnel, but the opportunity was lost; the controller population is now rising toward the original numbers.

Among those engineers who have worked in aviation

safety, the FAA has long had a reputation for technological backwardness. Cynics have been known to believe that the motivation is to preserve controller jobs, while others believe it is just inertia. Whatever the reason, every major study in the recent past—and there have been many— has concluded that the FAA facilities are far behind the technical state of the art, and that the rate of progress is so slow that the gap keeps widening. Those problems cannot be solved by hiring more controllers. And money is no problem—the Aviation Trust Fund is awash with uncommitted resources. Congressmen fly, and are eager to support any improvements in the system, provided the changes don't threaten their privileged close-in parking at the Washington airports. (Parking seems to be one of the stronger human drives. A former president of a large university, asked to describe his job, said it was to provide football for the alumni, sex for the undergraduates, and parking for the faculty.)

One simple measure of the FAA preoccupation with the role of the controller—FAA management is flooded with former controllers, since there is no other upward mobility in the system—is in the pay scales. A technician's highest pay grade, on the government scales, is GS-12 (about $42,000 in 1988), while that of a controller is GS-14 (about $60,000 in 1988). The arrangements for retirement are comparably tilted.

So far we have not mentioned the radar systems, approach systems, communication systems, computer systems, etc., so the judgment of technical inertia may seem unduly harsh. All these technologies find their places in the air traffic control system, but none, literally none, is even close to the capabilities of modern technology. Just as a single example, at a time when most commercial television broadcasts and private telephone conversations have long been carried by satellite—satellite relays avoid the problems of the earth's roundness—the FAA has no serious satellite plans. Instead, the control of each aircraft is transferred to a different controller, on a different radio frequency, every hundred or two hundred miles of its trip across the country, just as it was forty years ago. Elaborate procedures are used to coordinate the different Centers, and the FAA and its controllers deserve credit for doing as well as they do.

The navigation and communication system that services all this is marvelously intricate, and one marvels that it works at all. The major facilities include over a thousand separate very-high-frequency (VHF) enroute navigation facilities (of a type that was introduced forty years ago), another thousand or so lower frequency beacons (of a type that was introduced long before that, and is in fact not much used nowadays), twenty-five central Air Route Traffic Control Centers (where all the shuffling, computing, and coordination of the enroute traffic occurs), nearly seven hundred control towers to steer traffic into and out of airports, over a thousand instrument landing systems (also about forty years behind the state of the art, but in this case effective and reliable, and therefore candidates for replacement), thousands of radios, operating on a thousand frequencies, lots of radar (outmoded, of course) to provide information to the system, lots of computers (also woefully ancient, prone to breakdown, and slowly being replaced by slightly less outdated designs), and an assortment of other equipment, all presided over by tens of thousands of controllers. All the radio and radar facilities are on the ground, and therefore limited by the line of sight, so an IFR flight across the country consists of threading one's way through this maze of helpful electronics and people.

All things considered, the system works reasonably well, but how can that be? After all, one shouldn't quarrel with success, but should try to understand it. We begin by asking just how crowded the skies really are.

On the average, there are 1100 airline aircraft aloft at any given time (1986 data, and increasing), and 3300 general aviation aircraft (also 1986, and decreasing). If we add a few hundred commuter and air taxi aircraft, that makes a total of nearly 5000. The aircraft are unequally distributed, more in the East than the West, more low than high, more in the day than the night, and more in good weather than bad; the average will do for our purposes. Since the continental United States has an area of over three million square miles, there would be over six hundred square miles for each airplane if they were evenly distributed. They are of course spread out in altitude from the ground to a few tens of thousands of feet, so they wouldn't exactly be rubbing wings, even if no pilots were looking outside. Of course, a pilot is *supposed* to be looking outside in good weather, which is well over 90 percent of the time. That is the environment of the air, except in the vicinity of large busy airports and in other parts of the East, where the sky is really and truly full of airplanes. There, especially in bad weather, is where the ATC system earns its keep. On the average, however, the controller doesn't contribute much to the avoidance of in-flight collisions; the skies are normally not very crowded.

Indeed, the problem of collision avoidance is somewhat exacerbated by the existence of the system. Though the sky is big, and the world we live in has three dimensions, controlled aircraft are normally required to fly at altitudes that are multiples of a thousand feet, like 10,000 feet, 28,000 feet, etc., so they are crowded together at specific altitudes, and the potential for collision is increased. In addition, they are usually required to fly on specific airways (highways in the sky), which also crowds them. The reasoning behind this is that a controller who knows the exact location of each airplane can provide the necessary separation. It works most of the time, but is made harder by the channeling. The system works to put all the airplanes close together, to facilitate tracking, then works harder to keep them from colliding. Except near large airports, random flying would do nearly as well.

Technological advances are slow in coming. Procedural changes are not so slow, because they require only a signature, and they almost always consist of extensions of FAA-controlled airspace, and division of it into mind-boggling fragments.

In 1987, an Air Mexico jetliner collided with a small aircraft in a segment of controlled airspace near Los Angeles International Airport. The small aircraft shouldn't have been there. That segment of airspace was exactly 1,000 feet thick, extending from 6,000 feet to 7,000 feet altitude, with uncontrolled airspace (for these purposes) both above and below, to the north, to the southwest, and to the east. It was approximately pie-shaped, and it was literally impossible to determine from the navigational aids available in the sky, in either airplane, whether one was inside it or outside. (It was clearly marked on the map, but that is no help if the boundaries can't be defined by the navigational radios. One boundary was defined by tuning a radio to the Ontario beacon, another by tuning to Seal Beach, and the third by tuning to Los Angeles. Nobody has that many radios to spare. Besides, it is customary to devote some radios and some attention to the humdrum business of landing at the correct airport.) This doesn't excuse the pilot who was in restricted airspace, but does provide a partial explanation. It was a small chunk of oddly-shaped controlled airspace in a surrounding uncontrolled sky. For the airline aircraft, that is no problem; the controller tells the pilot where to go. Since he is cleared into the restricted volume, he doesn't need to know whether or not he is actually in it—the controller knows where he is. Such complex zonings of the sky exist around all the major airports, and are called terminal control areas, TCAs. The Los Angeles TCA has twelve of these oddly shaped segments, each with its own altitude restrictions.

The FAA response to the accident was to further restrict the airspace, make it more complex, and enlarge the controlled volume (among other changes, raising the top of the TCA from 7,000 feet to 12,500 feet, though the accident occurred below 7,000 feet). It then closed off the existing routes (the so-called VFR corridors through the TCA) that were placed there in the beginning just to avoid this kind of accident. The panic response was so palpably counterproductive that the FAA had to rescind the last part. The Administrator of that time deserves credit for having backed off, even though he deserves no credit for having acted so thoughtlessly in the first place, nor for having retained the other ill-advised changes. The FAA response had little relevance to the root cause of the accident, but relieved a long-standing itch for more controlled airspace. In the special case of Los Angeles, most safety experts have long been recommending an alternate approach to airspace control through dedicated approach corridors from the east. (Since the Pacific Ocean is to the west, there is less traffic from that direction.) Unfortunately, though such an arrangement is very probably safer than the complex TCA, it leads to less controlled airspace rather than more, and has attracted no interest at FAA management levels. Many regulatory agencies honestly believe that more control is synonymous with more safety. In fairness, they are not always wrong.

Aviation safety is in pretty good shape, but not just because of regulation. For the airlines, a real contributor is their own recognition that the flying public is sensitive to the occasional accidents that do occur, and that it is just bad business if there are too many. Pilots also know that their own lives and careers are at risk every time they fly, so they are generally a safety-conscious and professional group. Still, the old saying that flying is hours of boredom interspersed with moments of panic remains true. The National Transportation Safety Board (NTSB) has estimated that 43 percent of fatal accidents in commercial jetliners are initiated by pilot error.

Another major contributor to safety is the NTSB itself. An airplane is a complicated device, subject to both expected and unexpected stresses, and so is the pilot. No engineer could sit down at a drawing board to design out of the whole cloth a safe, stable, all-weather, fault-tolerant airplane. Nor could anyone wake up with a vision of the complex system of communications, radar, lighting, runways, signals, etc., within which the airplane must operate. Nor would it be possible to invent just the right kind of training for pilots, engineers, technicians, and maintenance personnel, let alone the controllers themselves. All this has to be learned from experience.

Today's airplanes don't differ in concept from those of eighty years ago, when the article quoted at the beginning of the chapter was written. There are dramatic differences in engines, structures, knowledge of the theory of flight, control systems, and so on, but the similarities would be more impressive to an extraterrestrial observer than the differences. Nothing is wrong with that—it provides a learning opportunity. Santayana was quoted earlier as having said that those who cannot remember the past are condemned to repeat it; that is especially true of accidents. Much of the activity surrounding a transportation accident is devoted to fixing the blame, but the NTSB investigation is directed toward learning from the mistakes. Its findings cannot be used in lawsuits.

The NTSB was not always independent, having been born in 1966 subject to the budgetary control of the Department of Transportation. True independence came in 1975, with the belated recognition that budgetary control is *de facto* control. When you hear on the television of a large aircraft accident, the story will usually include the statement that federal investigators are on their way to the scene. Later you will hear that the "probable cause" of the accident was determined to be pilot error, or continued flight into adverse meteorological conditions, or engine failure caused by fuel starvation (polite talk for running out of gas), or something like that. Such information comes from the NTSB.

The NTSB is not a regulatory organization, and has no operational responsibility. It makes recommendations, which the regulatory agency—in the case of aviation, the FAA—can accept or reject. Though there have been times when it has been politicized (the members are presidential appointees), it is widely regarded as effective and impartial. Certainly its reports are models of fact-finding probity.

The central function of the NTSB—there are others—is to force the system to learn from experience. For each accident for which a root cause can be determined, recommendations are made for reducing the chance of recurrence. These recommendations may involve rule changes, modification or inspection of aircraft, improved or modified training for pilots or non-flight personnel, changes in manuals, or just warnings to be careful about something. It is vital that the NTSB be independent, to preclude allegiance to any participants in the accident. Even with all the honesty and goodwill in the world, that would be too much to expect of the FAA—the designer, inspector, overseer, and operator of the air traffic control system. The FAA may have contributed to the accident.

So it is possible, over a period of time, to find out what is wrong with the technology and the system, by recognizing that it will sometimes fail, but not letting it fail in the same way over and over again. In the long run, the weak spots get patched. This process has led to an extremely strong aviation safety establishment, drawing strength in a systematic way from past experience; it is a model worth copying. Every organization will tell you it learns from experience, but one learns from experience to be suspicious of such claims. Santayana has already been quoted, so now is the time for Patrick Henry, who said, "I know no way of judging of the future but by the past."

But independence is a mixed blessing. Just because the NTSB, the investigator, is independent of the FAA, the operator, it is one step further from effectiveness. Its only influence, if the FAA disagrees with a conclusion or recommendation, is through persuasion or publicity. Before independence, however, unpleasant conclusions could be more easily swept under the rug. Consider the famous incidents of the DC-10 baggage door.

In the DC-10, as in other large passenger jet aircraft,

the body is divided longitudinally into two major sections, the passengers in the top section and the baggage down below. The floor of the passenger cabin serves as the roof of the baggage compartment. The floor also serves as a support for many of the flight and engine control lines that connect the cockpit with the tail sections of the aircraft. (The DC-10 has two engines mounted on the wings and one on the vertical fin.) The baggage compartment is sealed, and meant to be at approximately the same pressure as the passenger compartment. As the aircraft gains altitude the outside atmospheric pressure decreases, but passenger comfort and survival require that the inside pressure not decrease as rapidly, and it is customary to maintain cabin pressure close to sea level, even at very high altitudes. So the baggage and passenger compartments are at the same pressure, higher than the outside air, and a floor separates them. Any major loss of pressure in either will therefore stress that floor. The ground crew must close and secure the baggage door before flight, while the flight crew takes care of the passenger doors. There are indicator lights in the cockpit to report if anything is amiss.

In early 1972 an American Airlines DC-10 departed Detroit en route to Buffalo, and had climbed to about 12,000 feet when the baggage door blew off, leading to rapid decompression of the baggage compartment. Because there was now more air pressure in the passenger compartment than in the baggage compartment, the floor between them buckled. This compromised some important control systems, causing a major loss of control. Fortunately the aircraft was lightly loaded, the loss of control was partial, and the pilot was able to return the aircraft to Detroit. The airplane was badly damaged, but no one was killed.

Of course the NTSB conducted a full investigation, and concluded that the latching mechanism for the baggage door was poorly designed; it was possible to force it into place by brute force, without actually locking the door. The details of the locking mechanism are unimportant, except to note that forcing the door closed also turns off the cockpit warning light. Though the ground crew had noticed that the door was hard to close, they apparently didn't mind using force, and the accident thereupon became unavoidable. The flight crew never knew the door wasn't properly latched.

There followed a major conflict between the NTSB and the FAA about what should be done. The NTSB recommended that the FAA require a modification of the latching mechanism to make it physically impossible to jam the door shut in an unlocked position. They also recommended relief vents between the passenger compartment and the baggage compartment, to avoid the pressure differential that had caused the structural damage. There is no point in dwelling on the personalities involved, though subsequent events separated the black hats from the white hats. The FAA declined to require major modifications. Instead a small window was ordered installed on DC-10s, so the ground crew could see the latch from the outside, to make sure it was properly engaged.

In 1974, two years later, a Turkish Airlines DC-10 took off from Paris, and gained an altitude of about 12,000 feet before the baggage door blew out. This time the aircraft was full, the collapse of the floor was more complete, and so was the loss of control. It was the worst aircraft accident in history, and 346 people died. Almost identical to the earlier one except for the outcome, it was completely preventable. The inspection window had been installed, but apparently no one had told the ground crew to use it. It is well off the ground and hard to use anyway.

All DC-10s have now been properly modified, and in 1975 Congress made the NTSB a fully independent agency, but not before it had lost some of its better people. There is merit in independence, especially to enforce learning from experience; oxen may need to be gored.

Small Aircraft

The term "small aircraft" is sometimes regarded as synonymous with general aviation, but it is not. When publishers of titillating magazines travel around in their private DC-9s, that is part of general aviation because it is outside the airline industry. But the vast majority of general aviation aircraft are in fact small, carrying fewer than a dozen passengers, most commonly two or four. The two-place airplane is most often used for training, with the four-place one serving as the workhorse of the community. A stroll around the local airport will convince anyone of this.

There are over 200,000 of these aircraft in the country, and we have said that they make over 90 percent of all takeoffs and landings, carrying many fewer passengers than the airlines over shorter distances. Despite the widespread impression that much of this flying is for fun ("skylarking" is the term), the FAA classifies only about 6 percent as "recreational." It is a diverse collection of aircraft, operated by an inhomogeneous group of people for a variety of reasons, as befits a category whose definition is "all other."

The best general aviation aircraft are as well equipped with instruments and equipment as most commercial aircraft, while the least well equipped may be confined to a compass, an altimeter, and an airspeed indicator. It doesn't take much equipment to do crop dusting, and crop dusters don't usually fly into large airports.

The pilots vary greatly in skill. Of the 700,000 licensed pilots (only about 40,000 women), 250,000 hold the instrument ratings that permit them to fly in bad weather. Nearly all commercial pilots have the rating, but only a tenth of the 300,000 private pilots. If the numbers don't seem to add up, it is because the grand total includes 150,000 student pilots, who may fly under the supervision of an instructor but may not carry passengers.

General aviation is a far more complex mix than commercial aviation. And the accident rate is high, compared even to motor vehicle accidents. The number of fatalities has decreased rather sharply in recent years, dropping below eight hundred in 1987, but partly because the number of hours flown has also gone down.

Detailed statistical analyses of the causes of the fatal accidents in general aviation are published by the NTSB, but useful generalizations are hard to find. Of the 426 fatal accidents in 1987, which killed 788 people, there were almost 426 causes. Such accidents tend to be categorized more than they are analyzed, though any analysis may result in a modification order of some kind for other aircraft of the same type. To that extent the procedures are the same as they are for commercial aviation. The differences are that a commercial aviation accident is rarer, kills more people, is more thoroughly investigated, and constitutes a unique lesson in itself.

One can gain insight by looking at the broad categories used by the NTSB. It lists an event known as the "first occurrence," which is the initial event in the accident sequence. The leading causes are loss of control in flight and encounter with the weather in flight. Neither involves other aircraft or the air traffic control system, and both causes are entirely avoidable. This is reflected in the NTSB's assignment of probable cause, which lists the pilot as a factor in fully 90 percent of general aviation fatal accidents, with weather contributing to about 35 percent. The next largest category, which is not much help, is "miscellaneous." Pilots mishandling aircraft under adverse conditions, going into weather for which they are not qualified, bad (in retrospect) decision making, and generally poor judgment—those are the ingredients of which the fatal accident stew is made.

But there is none of the prevalence of alcohol or youth in these accidents that there is for automobile accidents. Although there is some alcohol involvement it is not the major influence it is on the road. Similarly, the age distribution of pilots involved in fatal accidents is reasonably evenly distributed between thirty and fifty-five, dropping off at both ends. It is not that the young are not allowed to fly—the age requirements are sixteen for a student license and seventeen for a full-fledged pilot's license—but that they don't have the easy access to aircraft that they have to cars. Doubtless there are also social factors.

Nor are the pilots involved in fatal accidents necessarily inexperienced. Pilots with thousands of hours at the controls get into situations they can't handle. To be sure, there is a period just after about a hundred hours of experience in which a pilot thinks he knows everything there is to know, but he will be a much better and more humble pilot if he survives his first encounter with his limitations. In the same vein, about half the fighter pilots who were killed in combat in World War II were killed on their first combat mission.

There is not much more to add. The source of most
general aviation fatalities is not in the technology, the airplane, or the air traffic control system. It is in the people.
And that is too often true of large aircraft.

Large Aircraft

The air traffic control system is geared to service the large
aircraft so familiar to most Americans. There are about
three thousand of these, and airline economics requires
that they be in the air about a fourth of the time. The
more the better, since an airplane on the ground ties up
capital but earns no revenue. As we said at the beginning,
the major service the control system provides is separation,
especially in bad weather. In good weather, most pilots are
vigilant, most of the time. Yes, there is some laxity and
complacency at high altitudes, when all aircraft are under
positive control, separated in altitude, and there aren't
many of them anyway, but midair collisions are rarities.
Airline accidents, like general aviation accidents, often follow from pilot error. To be sure, generalizations are hard
to draw, since they are so few, each is a special case, and
each is a complex story in itself.

It is a rare accident that is a single event, an aircraft
cruising through the sky, its occupants carefree and happy,
when disaster strikes without warning. Careful study of
the root causes will almost always show that the accident
had been waiting to happen, that it had signalled this fact
through small earlier breakdowns or in other ways, and
that it was always preventable in retrospect. Retrospect
is a great tool, but it is never there when you really need
it, in advance. When pilot error is singled out as the major cause of an accident, that usually means that the pilot
dealt badly with some abnormal event, whether a small or

large mechanical failure, or weather, or any of the dozens of challenges he would normally take in stride. Human error also includes the maintenance and other support personnel whose collective efforts make for a reasonably safe flight. The pilot is supposed to cope with all the remaining problems, but he is at the end of a chain with many links. His training is in dealing with abnormalities—any dunderhead can learn to fly a modern airplane under routine conditions.

Aircraft and other systems for which safety is important tend to rely on some general safety principles that span all the technologies. The first is redundancy—most vital components are duplicated or triplicated, so that the failure of one will not compromise the system. That is one reason for airline aircraft to have two or more engines, and to be designed so that loss of a single engine leaves the aircraft flyable. Pilots practice engine-out procedures regularly. Instruments are duplicated, radios are duplicated, even pilots are duplicated, so as to make the aircraft resistant to what are called single-point failures. When a pilot claims that all the engines failed in flight (on a multi-engine airplane), most other pilots assume he made a mistake, or ran out of fuel. A familiar mistake in twin-engine aircraft is to suffer an engine failure, and then respond by shutting down the good engine. It is not always easy to know which engine has failed. But that sort of embarrassing error is unusual, and redundancy goes a long way to assure the safety of aircraft.

When a component can't be duplicated, the FAA standards for aircraft design require that the chance of failure be "extremely remote," though the interpretation of that requirement in terms of probability is unclear. Right wings, for example, are not duplicated on modern aircraft, so they should be unusually reliable, and indeed structural failure in flight is rare.

The second safety principle that is ubiquitous in design is called defense in depth. In the event that something fails, there ought to be some backup system that will either provide the same function in another way, or at least provide a safe way out of the predicament caused by the first failure. There is always another way to lower the landing gear if the normal method fails. There are mechanically and electrically driven pumps. Some limited communication is available through the radar transponder, if all radio communication is lost. Pilots actually remember how to fly the airplane if the autopilot fails.

The combination of these principles, when coupled with reasonably conservative design, makes large commercial aircraft pretty safe machines, and that is reflected in the record. (Of course, conservatism can lead to false illusions of improved safety, as was emphasized in Chapter 10.)

Though it is not normal practice, one can do probabilistic risk analyses (PRA, as described in Chapter 5) on aircraft, and quickly learn that the provisions of redundancy and protection against single-point failures, combined with defense in depth, make the calculated failure probability low indeed. One can also learn that the remaining risk is in the area least amenable to PRA, so-called common-cause failures, also mentioned in Chapter 5. It is very difficult, verging on impossible, to predict the probability that some event or human act will compromise all the elaborate redundancy and defense in depth that are built into a well-designed aircraft, but experience shows that it does happen. Here is one famous example, which by pure good fortune killed no one.

It is normally considered good to have two engines instead of one, though a few dissenters point out that that doubles the chance of an engine failure in flight. How much better it must be to have three or four. The probability that any single engine may fail is known, and the probability that two of them fail *independently* can also be calculated; it is truly minuscule. But *independently* is the key word, and it does happen.

On May 5, 1983, a Lockheed L-1011 airplane (with three engines) was on a short flight from Miami to the Bahamas, when the crew noticed a loss of oil pressure in one of the engines, and shut it down. That was the correct procedure to avoid damage to the engine, since the airplane is eminently flyable on any two engines. (When the oil pressure idiot light in our cars goes on, we should also shut down our engines.) The captain decided to return to Miami, for its better facilities, again a reasonable decision. Fifteen minutes later, after they had reversed course and were cruising toward an expected uneventful return, another engine failed, and five minutes after that, the last. That left the aircraft functioning as a glider, for which it was not designed. They were over water, without power, losing altitude, and in real trouble.

At about 5,000 feet, having prepared the 162 passengers for imminent ditching in the ocean, the crew was able to restart the engine that had first failed, and it was barely enough to get them to Miami. Happy ending, but how can all three independent engines on a modern airplane fail within minutes of each other? The usual cause of such an event is what is euphemistically called fuel mismanagement—running out of fuel, which would be gross pilot error—but that wasn't the case here. The flight crew came out completely clean.

The three engines had run out of oil, not fuel. But how can that happen?—they have independent oil reservoirs, and all had been checked before takeoff. Of course that was the problem, they had been checked. Whenever something is checked, there is a chance of leaving it in worse shape than before. Part of the routine maintenance of this type of aircraft was and is periodic removal of magnetic chip detectors, little magnets that are inserted in the oil lines. They are there to attract and hold any stray particles of steel, which would be indicative of undue engine wear or damage. They are there for safety reasons. Each has two O-rings on it, to prevent oil leakage. Unfortunately, O-rings work only if they are installed, and none of the three chip detectors had been installed with its O-rings in place, so all three engines leaked their oil and bled to death. This was a classic common-cause failure, the common cause being the same maintenance crew. It is easy to prevent; don't let the same crew get its hands on two elements of a redundant system. But that makes scheduling problems, and is rarely the practice.

In this case, the story is even more depressing; the problem had been crying for attention. In the year and a half before this event there had been twelve separate occasions on which engines of this type had suffered oil loss, nine of them resulting in shutdown of the afflicted engine, and five of those were due to installation of master chip detectors without their O-rings. The FAA inspectors knew about all this, but didn't see that it could happen some day to all the engines on a single airplane. They contented themselves with telling the airline to be more careful and to change the maintenance work card.

In the next section we'll talk about some more famous O-rings, but the story there is also one of not taking early failures very seriously. It seems that we believe we are leading charmed lives until facts to the contrary are literally jammed down our throats.

All fatal airline accidents, and many that are not fatal, get the full NTSB treatment, resulting in detailed reports of the circumstances that led to the accident, the sequence of events, and the assignment of principal and contributing causes. This was no exception, and what is written above is based on the NTSB report. The reports are intended to make it as painless as possible to learn from experience, by providing advice to the community. The best way to understand the complexities of aviation safety is to read a few of these reports.

Spacecraft

Strictly speaking, spacecraft don't belong in a chapter on air transportation—space is almost free of air—but the safety issues follow naturally from what has gone before. The industry that supports the space program is the same as the one that supports aviation, and the engineers call themselves aerospace engineers. Much of the technology is the same. The design challenges for space vehicles are more stressful than they are for aircraft, but the safety issues are handled more clumsily, at greater expense, with fewer vehicles. All in all, the similarities justify the juxtaposition of aircraft and spacecraft.

Before the Challenger disaster in January of 1986, the United States had flown twenty-four orbital missions of the space shuttle, and a comparable number of pre-shuttle flights, some of which went beyond earth orbit. There had been a fatal accident on the pad almost exactly nineteen years earlier, killing three astronauts, but no fatalities in flight. Of course there had been problems, malfunctions, and even drama, but no worse. It was a marvelous record, and the Soviets have done comparably. It was that record of success which led inevitably to the systemwide complacency that was the root cause of the Challenger event, and will be the root cause of the next one. Though the infamous O-rings were the agents of the tragedy, complacency was the villain behind the scenes.

The first object in orbit (Sputnik) was launched by the Soviets in 1957, followed by a dog (Laika), and finally the first manned orbital flight of Yuri Gagarin in 1961. (Assuming again that those who were under ten years old at the time don't remember any of this, that is more than half our present population.) The United States was not far behind—just under a year later, John Glenn (now Senator Glenn) orbited the earth. In the years from 1957 through the early 1960s the country was in a kind of intellectual panic, centered on the fear that we were losing the technological initiative to the Soviets. Everyone and everything was blamed, especially our educational system. The slogan "When in trouble or in doubt, run in circles, scream and shout" became popular. Naturally the nation marshaled its resources to "catch up," and in 1961 President Kennedy announced a program to land men on the moon, *and bring them back safely* by the end of the decade. So the Apollo program was born. The reason for repeating this bit of history is that it reflects the national sense of urgency and commitment that fueled the space program in those days.

The program was successful; the first lunar landing was made on schedule in July of 1969. Over the next three and a half years there were five more lunar landings and safe returns, one high drama involving an aborted effort in which the astronauts were able to circumnavigate the moon and return to earth, and no one has been back since. The Soviets never tried. Whatever the reason, we have concentrated ever since on unmanned exploration of the solar system, and use astronauts only in earth orbit. Many wonder why we do even that, since their functions on these flights are not uniquely human.

That was not the outcome envisaged in the heady days of the 1960s. In February of 1967, more than two years before the lunar landing, the President's Science Advisory Committee (PSAC) issued an influential report entitled "The Space Program in the Post-Apollo Period," outlining a recommended program of space exploration after the (anticipated) success of the Apollo effort. The suggested program included extended manned exploration of the moon, development of a manned space station, and a vigorous program aimed at manned exploration of the planets. All visionary, but indicative of the enthusiasm that brought out the best in all of us. None of it has happened; there is nothing remotely exploratory in the use of astronauts on shuttle missions. And it is a splendid group of people, fit for space exploration instead of chauffeur duty.

Even in the PSAC report there was a recommendation that "NASA study the advantages of adopting a planning and decision-making structure which emphasizes program objectives rather than the means used to attain them." That is, unfortunately, still a good recommendation, but not the subject of this book. It is nice to extend human presence in space—it would be even nicer to have a clear reason for doing so. In early 1989 NASA announced that it is going to ignore distractions while it "devotes its attention to proving the shuttle system's capability for safe, reliable operation." Sadly, that too is a means, not an end—the means have once again turned into the objective. Einstein said of his time, "Perfection of means and confusion of ends seem to characterize our age." *Plus ça change, plus c'est la même chose.*

The Apollo vehicles were the product of the aircraft industry, transmogrified into the aerospace industry, and the best skills available were brought to bear on their creation. The design, certification, and testing procedures were those that had worked in aircraft design, suitably modified for the novel application. And it all came together and worked reasonably well.

Early efforts were made to incorporate innovative risk assessment techniques like probabilistic risk assessment, but the program managers had little use for the calculated probability of launch success. One has some sympathy— does it do any good to say to an astronaut that the launch is approved because the odds against failure are three to one? And the computed odds were worse than that. As a result the only possible way to make quantitative assessments of safety was lost, and is still missing. You can get away with this in the aircraft industry because there are many aircraft and many flights, so experience alone provides enough successes and failures to point the way. For Apollo, with only a dozen flights, overall flight success experience reveals little about real safety. Like someone playing Russian roulette, if each success makes you more confident that you are going to live forever, you have fooled yourself and will pay the price.

So two pervasive safety problems crept into the system as NASA moved from the Apollo period into the shuttle era. The first was (and it hurts to say this) that the national enthusiasm faded, and with it the incentive for our finest talents to become involved. This is not specific to

NASA—there is a natural life to innovative programs. At the beginning, as new horizons are explored, visibility is high and rewards in terms of personal achievement and fulfillment are great—then the quality of the people attracted to the program is the highest the nation can provide. The most precious commodity the nation has is fine people. It is easy to get a reaction from an audience by pointing out that half the American people have below-average intelligence, yet it is true by definition (oh well, median, to the purists). That is not social commentary, but an indisputable mathematical fact. At the higher levels of capability there are even fewer members, and there is less incentive for those few to go into a program, or even stay in it, once it has lost its direction.

By widespread agreement this happened to NASA as it made the transition from Apollo to the shuttle. The problem is compounded by the fact that the nation as a whole has been losing ground to the rest of the world in mathematics and science. In 1988 the Educational Testing Service conducted a study in which eighth-graders from the United States, a number of other countries, and four Canadian Provinces, were tested in science and mathematics. Of course, the United States came out dead last. The educational crisis is real and devastating.

It seems to be hard to adjust to the fact that every person can't be above average. This author has served on many committees in which a proposed recommendation (always to solve some problem) was that somehow the people charged with doing the job should be made wiser than they actually are. Such recommendations don't help. He once chaired a committee on Air Force officer training and educational criteria, and learned that 90 percent of Air Force officers were rated "well above average," though that is a mathematical impossibility. (The other Services do no better.) He once heard a distinguished attorney say, after the accident at Three Mile Island, that the trouble with the nuclear industry was that too many of the utilities who own nuclear reactors are below average. When the attorney was reminded that, whatever we do, there will always be a worst utility, he replied that, in this business, we can't afford that. For every class, there will be a student with the lowest grades, and fully half the class will be below the median. It seems silly to say it, but only 10 percent of the students will be in the upper 10 percent. This simple fact, that there is a shortage of the kind of extremely high-quality people needed to make important and innovative programs work, seems to be one we don't have the courage to face. It smacks of elitism, and so it should.

The transition is particularly damaging when a program, like the space program of the 1960s, has been successful. That breeds complacency, and a sense of being charmed. It is never overt, and will always be vehemently denied, but it is almost inevitable. Without real demonstration of risk, through accidents, it is easy to believe they are impossible, and even easier to take the credit. Few have the character to say that a program is running on borrowed time because there were giants in the old days. Sometimes it isn't true, but often it is. Even before the end of the Apollo era, at the time of the fire that killed three astronauts on the pad in 1967, there was evidence that NASA was mismanaging its safety programs. Congress then created a new safety advisory committee for NASA, to provide a counterbalance to the complacency that was already developing within the agency. NASA didn't want it, and the Rogers (Presidential) Commission report on the Challenger disaster has a chapter entitled "The Silent Safety Program." The title speaks for itself, but, for reasons that are incomprehensible to this author, that doesn't seem to embarrass any of the participants, who hold themselves blameless.

Aaron Wildavsky has emphasized the point that all organisms need a certain level of risk in their environment, lest they drift further and further out of step with it, and are ultimately unable to meet its challenges. That is also true of programs.

Finally, when dealing with rare accidents, there are two major tools available to postpone the inevitable (apart from those that go without saying, like meticulous engineering, testing, design, maintenance, training, operation, and quality control). One of these is a systematic and energetic effort to learn from experience, rather than sweep unpleasant experience under the rug. The NTSB is responsible for bringing that positive element into the improvement of aircraft safety—there is nothing comparable at NASA. This too is denied, and it is said that it is everyone's responsibility to learn from experience, but that is another way to say it is nobody's. The fateful O-rings that failed in the Challenger disaster had failed many times before. Since they hadn't led to disaster on those occasions, the easy conclusion was that such failures were "within the experience base," and therefore acceptable. The problem of complacency has not yet been solved at NASA, and may not be until there is another tragedy.

The second major tool available for understanding rare accidents is analysis, specifically probabilistic risk assessment. It is, as described in Chapter 5, a systematic procedure for putting together everything known about component failure rates, systems interactions, and human errors, to provide an estimate of the probability of an accident. It isn't perfect, but is far better than omphaloskepsis (contemplating the navel, for those few readers who may not be familiar with the word). In particular, the fact that the resulting probability is always finite can have a salutary effect on those who prefer to believe, in their hearts, that an accident is impossible. Since the probability is truly finite, not zero, an accident *will* happen; the only important question is when. That is a good state of mind for people charged with safety responsibility. Probability is a natural enemy of complacency.

In the early days of the Apollo program, NASA was regarded as a national leader in the use of probabilistic risk assessment, but the probabilities for failure that it produced were uncomfortably high, and the methodology was banished from the scene. (Behead the messenger who brings bad news.) Thus, as Richard Feynman (one of the authentic geniuses of our time) says in his book *What Do You Care What Other People Think?*, it was possible for the NASA engineers to believe that there was almost no chance for the solid rocket boosters to fail, despite overwhelming statistical evidence that such rockets fail a few percent of the time. NASA still has an inexplicable mindset against quantitative risk assessment (this author has been excommunicated for recommending it), but the handwriting is on the wall. One can only hope that it will not absolutely require a new and dreadful accident to force improvement in the risk management system. Recall that assessment is not the same as management, but is an essential tool for monitoring the effectiveness of management.

This section has concentrated on NASA's risk management system because that is where the problems lie. The technology is challenging but within reach, while the management of that technology for safety is not up to the mark.

The memories of the Challenger accident are still so fresh that there is no point in recalling the sequence of events. Early warnings that the O-rings were prone to failure, especially in cold weather, were rationalized away; people who were concerned were told to forget they were engineers, and to act like managers; the burden of proof was placed on the cautious, not the intrepid; and the top NASA management didn't even know what was going on. So seven people who needn't have died, did. NASA's management blindness to the clear risk of space flight had been previously demonstrated in its willingness to carry civilian passengers, a senator, a congressman, a Saudi prince, and on the fateful flight even a schoolteacher. The story is well told in the Rogers Commission report, which spells out the supporting information for many of the sweeping generalizations made here. Feynman's iconoclastic book is good reading, and more than half of it is devoted to his experiences as a member (and apparently an irritating one) of the Rogers Commission. It would be nice to be able to end this section by saying that the shock of the first loss of American astronauts in space has led to major improvements in NASA's approach to risk, but it wouldn't be true. Most of NASA believes that the Rogers Commission report was egregiously unfair. It wasn't.

15

Ionizing Radiation

FAMILY CHARACTERISTICS

Ionizing radiation in large doses is bad for us unless we are getting radiation treatment for some disease; in that case the benefits outweigh the risks. It may or may not be bad in small doses—no one knows. Anyway, it is everywhere and unavoidable. What is it?

Each of the atoms from which we and our surroundings are made consists of a positively charged nucleus, surrounded by negatively charged electrons. When the negative charge in the electrons exactly balances the positive charge in the nucleus, the atom is called neutral. When they don't balance—too few or too many electrons—the atom is called an ion. Neutrality is the norm, but external influences can disturb the peace and tranquility of neutrality, just as in politics. Sometimes radiation passing through us can detach an electron from its host nucleus, creating an ion; the radiation is then called ionizing radiation, and there are many forms. X-rays are familiar,

218

gamma rays are not, cosmic rays come from the sky in great diversity, and would be familiar if we could see them. Any of these can ionize matter, including the matter in human bodies.

We are accustomed to medical and dental x-rays; they account for about a third of the average person's exposure to ionizing radiation. They help diagnose illness, but can also damage body cells, so we keep the exposure to the minimum necessary. That's the issue in the debate about whether regular mammograms are recommended for all women, or only for those at most risk. Similarly for chest x-rays and other medical procedures. Near the bottom of the virtue list is the x-ray exposure received from sitting in front of a color television set (the high-speed electrons that produce the picture on the screen also produce x-rays, which are reduced to "safe" levels, but not entirely blocked, by the thick glass in front.) In fairness, those "soft" x-rays don't penetrate very far into the human body. The damage is small and the social benefit debatable.

Medical x-rays are examples of voluntary exposure to radiation—two thirds of all exposure is involuntary, to natural radiation. Cosmic rays deliver about 20 percent of the dosage received by the average American, and cannot be escaped, even in the deepest cave. Their intensity is greater at higher altitude—about twice as much in Denver as at sea level—and commercial flight crews and passengers in high-altitude aircraft absorb even more. Air crews in the modern high-flying jets receive more radiation than is permitted for the general public, and as much as most radiation workers on the ground. Yet they come under no regulations, and are normally not even monitored for their exposure. Frequent air travelers can double their annual dosage of "normal" radiation. Astronauts get zapped even more, especially the Soviet astronauts who have spent more than a year in space. The ground also contains materials that emit ionizing radiation, as do our houses. Our own bodies emit radiation from which we obviously can't escape, mainly from potassium (an essential nutrient), inspiring the observation that it is safe to sleep with one person, but not with two. Because of the radiation.

That's not all. We and the Soviets stopped atmospheric testing of nuclear weapons in 1962, but left a residue of gradually decaying radioactivity, still detectable. Nuclear power plants normally release small amounts of radioactive materials; in an accident they might release more. Nuclear waste must be disposed of carefully. All air contains radioactive radon and its products. Cigarette smoke contains radioactive polonium. The glass used in eyeglasses has small quantities of radioactive uranium and thorium. So does crockery.

This must sound like a horror story, designed to make you fear radiation. But the vast majority of all these radiation sources deliver extremely small doses, with minimal (if any) health effects, even though fear of even trivial doses of radiation is common.

The unit in which radiation dosage is measured is the rem (for roentgen equivalent man, after the discoverer of x-rays), and a thousandth of it is called a millirem, abbreviated mr. Ignoring radon, which is a story in itself and comes later, the average American is exposed to about 150 mr per year, from a variety of sources (including sleeping partners). The maximum permitted exposure for workers in nuclear facilities is 5,000 mr per year, and for the general public 500. We don't know if this much radiation does any harm at all. There are even people who think that radiation, like wine, is healthy in small doses.

(As used here, the word "unknown" has a one-sided meaning. We know the radiation can't be doing much harm, but can't certify that it is doing no harm at all. Prudence requires us to assume that it is harmful, but that may be wrong.)

The 500 mr per year limit for the public is lower than the amount one can receive from a color television set emitting its quota of x-rays. Of course a person spending too much time in front of a set is more at risk of having his or her brain turned to mush by the programming, so the radiation may not matter much in the grand scheme of things.

All the direct information about the effects of radiation on humans comes from the few times that people have been heavily irradiated, whether accidentally or deliberately. There are, of course, also animal data, with the same caveats as in the chemical case. Switching back to rem as a unit (recall that one rem equals 1,000 millirem), we know that an exposure to about 400 rem will kill about half the people so exposed, though competent medical attention can help. This is known from the few people who have been accidentally irradiated and from the two nuclear weapons that were used in anger during World War II. The knowledge was confirmed at the Chernobyl nuclear accident in the Soviet Union, where thirty-one died.

At somewhat lower doses, perhaps 50 rem to 400 rem, people will suffer radiation sickness, with damage to internal organs, but will usually survive and recover. Below about 200 rem, survival is almost certain. Many at Hiroshima and Nagasaki survived such irradiation, as did about two hundred people at Chernobyl. Below about 50 rem, our bodies can handle it. That doesn't mean there are no effects, only that we move into a different regime, one in which obvious acute damage does not occur. The radiation may, however, have seeded or promoted a cancer that will show up in ten to fifty years. That is the effect most feared.

We know that moderate doses of radiation can cause cancer. Most of the evidence is from the atomic-bomb survivors, but there is more, like the medical use of radiation before the risk became clear. For example, ankylosing spondylitis, a spinal disease, has been treated with moderate doses of x-rays, as has a form of ringworm of the scalp, and cancer is more common in people so treated. But those dosages are in the range of 100 rem; no harm has ever been demonstrated at the much lower doses to which we are normally exposed. To repeat what has been said many times, that doesn't mean that there is no harm, only that any harm there may be is undetectable.

All regulatory issues fall into that low-dose regime where, despite lack of knowledge, decisions about the protection of the public must still be made. The Committee on the Biological Effects of Ionizing Radiation (BEIR) of the National Research Council has produced the most authoritative estimates, though not without internal squabbling. They have concluded, based on their collective expertise, that the probability of radiation-induced cancer is most likely a combination of a linear and a quadratic function of the dose. This means that the probability is not directly proportional to the amount of radiation, but somewhat less. This semi-official assessment is used in all responsible estimates of the effects of specific events, like Chernobyl. It is the best we have, and the committee has acknowledged explicitly that it is not known whether doses in the range of 100 mr are at all detrimental to man. The lack of knowledge actually extends to substantially higher levels of irradiation. (At this writing, the latest version of the BEIR report is just being released, and it adjusts upward, by a factor of about three, the previous estimate of the effects of low levels of radiation. This is based on a revision downward of the exposures to the Japanese victims in World War II. It is a dynamic subject.)

The reason this matters is that all the stories one reads in the press about the tens of thousands of Russians who are going to get cancer from Chernobyl in the next fifty years are based on these estimates. There may be none at all, but that also won't be known. Ionizing radiation is neither benign nor the fearful monster it is often made out to be.

The benefits of ionizing radiation have hardly been mentioned, because this is a book about risk, but radiation has such a bad name that it would be only fair. Many of the uses of x-rays are familiar. They make the diagnosis of illness and the setting of broken bones less of an art form, and less chancy, than they once were. Medical diagnosis through radioactive tracers is now common, as is radiation therapy for cancer, where it sometimes cures. Radioactive iodine, which collects in the thyroid gland, is used to treat both benign and malignant thyroid disease. Medical research uses radioactive tracers routinely. There is less radium used on watch dials than there used to be, but more tritium (a radioactive form of hydrogen) is used for lighting. Radioactive sources are used to study metal components for critical structures, and to supply power for space exploration. Radioactive dating is essential to archæology, geology, anthropology, and a host of other –ologies. We'll get to nuclear power shortly.

But the most bizarre use of radioactivity is in the testing of alcoholic beverages for authenticity. Ethyl alcohol, the *raison d'être* of alcoholic beverages, can be made in many ways. The traditional way is through the fermentation of grain, grapes, honey, sugar, or other carbon-containing foodstuffs, but alcohol can also be synthesized from minerals like petroleum. Æsthetes would prefer the former to the latter (petroleum wine?), but how can a person tell? Alcohol is alcohol, and our bodies don't know the difference. Still, it's the principle of the thing.

Carbon comes in several forms, called isotopes (nuclei of different weight), all but two of them radioactive. The radioactive isotopes have short lives (see below for a discussion of the half-lives of radioactive materials), so a very old sample of carbon has only the stable isotopes, the others having long since decayed. Coal and petroleum are very old—they have hardly any of the radioactive isotopes. In the atmosphere, however, the constant bombardment by cosmic rays produces nuclear reactions that replenish the supply of the major radioactive isotope, carbon-14 (C^{14}). Therefore anything made of plant materials, which have breathed the air and used the carbon dioxide in it for food, will have its fair share of C^{14}, whose later decay can be observed. (The half-life of C^{14} is over five thousand years, longer than anyone is likely to age even a fine bottle of wine.) So it is possible to tell through radioactivity whether a given alcoholic beverage was synthesized from minerals, or is the real McCoy. The good stuff is radioactive, while the *ersatz* is not! Government testing has revealed that a number of imported alcoholic beverages are insufficiently radioactive.

Nuclear Power

Even when so much of the scientific community was preoccupied with the development of atomic bombs in World

War II (the book *The Making of the Atomic Bomb,* by Richard Rhodes, is a marvelous account of that epic time), some were thinking of using this remarkable new source of energy to make electricity in the postwar world. It seemed an essentially limitless source. The scientific community has taken a bum rap for the 1954 statement that electricity might eventually be too cheap to meter, but that was never said by any scientist. Unlike scientists, politicians and admirals are not accountable for what they say, and even the person involved was misquoted out of context. In 1988, nuclear power supplied over 20 percent of our electricity; coal supplied more than half. Coal has its own collection of risks, which are reserved for the next chapter.

It was recognized at the beginning that nuclear power plants would concentrate an enormous amount of both power and radioactivity in one place, posing major safety problems. Any radioactivity escaping into the environment might do real damage, so the business of nuclear safety has always consisted of keeping the radioactivity in its proper place, even in an accident. Risk is implicit in the use of nuclear power, and the problem, just as for any other source of risk, is to keep it down to an acceptable level.

(Another disclaimer. This author serves on some government advisory committees dealing with nuclear safety, and has chaired several studies on the subject. The views that follow are personal, as elsewhere in the book, and shared by depressingly few. He is also a supporter of nuclear power, while an advocate of more focused attention to nuclear safety. He is therefore viewed by his pro-nuclear friends as too anti-nuclear, and by his anti-nuclear friends as too pro-nuclear. *C'est la vie.*)

Nuclear energy comes from the fission, or splitting, of

certain particularly susceptible nuclei, of which an isotope of uranium, uranium-235, and an isotope of plutonium, a man-made element, are best known. The so-called atomic age (a misnomer, since it is not atoms that are split) began with the discovery that these fissile nuclei could be caused to split roughly in half by bombardment with subatomic particles called neutrons; this releases enormous amounts of energy, which can then be used to make electricity. A nuclear generating station, in the most common current form, contains about a hundred tons of uranium oxide. The fission energy is used to boil water, which produces steam, which then turns turbines to generate electricity. After the steam is extracted, a nuclear power plant is not much different from a coal or oil or natural gas plant. Anti-nuclear buffs are fond of poking fun at nuclear energy by saying that it's just a fancy way to boil water, as if they knew how to boil it by waving a wand. As all engineers know, boiling water provides an extremely effective and efficient way to convert heat energy to other forms of energy. Pound for pound, uranium-235 packs about ten million times as much energy as coal.

The fissioned nuclei leave a residue of extremely radioactive fission fragments, and this is the radioactive material that must be kept out of the environment. Many cartoons at the times of Three Mile Island and Chernobyl—the two best-known accidents at this writing—showed a damaged reactor with a suggestive mushroom cloud over it. Yet it is absolutely and unequivocally scientifically impossible for a reactor to blow up like an atomic bomb. (That statement is as clear and unambiguous as possible, with a surfeit of adverbs, because the question keeps coming up.) Editorial cartoonists are also not held responsible for what they draw.

We are not trying to slight the risk—the amount of radioactivity held in the core of a functioning nuclear reactor is truly prodigious, by any previous standards, and is therefore potentially extremely dangerous. Extraordinary measures to keep it in its place are fully justified.

How can it get out? Routine emissions of radioactivity are kept small and pose essentially no risk, so accidents are the issue. The reactor has systems that (if functioning) can stop fission within the reactor, but heat continues to be generated by the nuclear fragments even after the fissioning process is completely stopped. If a reactor is operating at full power, has been doing so for a while, and is suddenly turned off, it will still generate about 7 percent of its previous power. That power, called decay heat, will slowly decrease over a period of hours, days, weeks, and months. It is a kind of coasting down, and is unavoidable. Even though 7 percent may seem like a small number, it can still be as much as two hundred million watts, enough to supply a small city. So reactors must be continuously cooled, even after they are turned off, or they will melt. If they do melt, it is called a core-melt accident, and is a potential disaster. At Three Mile Island, part of the core melted without disaster; at Chernobyl, the outcome was far worse.

Under normal operating conditions the fission reactions are turned on and off in an orderly way by the control systems. They can also be stopped quickly, automatically or by hand, if anything is abnormal. The special system that does this is called the SCRAM system, and the job is so important that it must be unusually reliable. (There is a lively debate among the old-timers in the nuclear community about whether the expression SCRAM is or is not an acronym for something. Since their memories differ, we may never know.) Though there have been partial failures of SCRAM systems, there had never, until Chernobyl, been a complete failure.

United States reactors average a few SCRAMs a year, somewhat fewer for the Japanese. A SCRAM is like a doughnut—you can concentrate on the good stuff or on the hole. Each SCRAM reflects proper functioning of a safety system, and that is good, but if the safety systems are challenged too often, that is bad. A failure to SCRAM when called for can be very serious, and the less often a SCRAM is needed the less often it is likely to fail. In the strange jargon of the community, the occasional but regular SCRAMs due to abnormalities are called "anticipated SCRAMs," because a certain rate of abnormality is unavoidable. There are also unanticipated SCRAMs, whose definition is "all other."

The most studied nuclear safety problem involves loss of cooling, whether through a broken pipe or some other breakdown in the reactor cooling system. If the reactor core is covered with water, and stays that way, melting can usually be avoided; otherwise there is a major problem. When coolant is lost from the reactor we have a loss-of-coolant accident, or LOCA. Such an accident is easy to visualize—it is a water leak—and was therefore the focus of attention by the anti-nuclear community during the 1970s, but it is not the most likely source of real trouble. The concentration of attention on LOCA led in the 1970s to neglect of other safety issues, and contributed to the accident at Three Mile Island.

How safe is the commercial nuclear enterprise, and where is it heading?

The art of probabilistic risk assessment is most advanced for nuclear power, so we can calculate the probability of a major nuclear accident tolerably well. That kind of "bottom-line" estimate is subject to great uncertainty, but is still better than guesswork. The best current estimates for American reactors are that the probability of a core-melt accident is about one in ten thousand, per reactor-year, and the chance of a major release of radioactivity to the environment is a tenth or a hundredth of that. If these numbers are taken seriously, a core melt would be expected about once every hundred years among the hundred-odd reactors now operating. There is no way to predict with any confidence whether or not there will be a core melt before the turn of the century, because the art of prediction isn't that good. Probably not, but no promises.

These numbers are consistent (which is no surprise) with the announced safety goals of the Nuclear Regulatory Commission, but should of course be taken with a grain of salt. The uncertainties are large. As time goes by, and both wisdom and experience accumulate, we will know more.

ACCIDENTS

Three Mile Island

The accident in 1979 at reactor Unit Two on Three Mile Island, a small island in the Susquehanna River, not far (but much more than three miles) from Harrisburg, caught the world's attention. It was the first ever major nuclear accident, was featured on the evening television every night, was headline news throughout the civilized world, and kept people hanging for a week. It also generated attendance for the movie *The China Syndrome*, which by a stroke of good luck for the producer depicted a similar event.

The reactor core was partially melted, the utility suffered financial disaster, the reputation of the NRC was properly and deservedly tarnished, and the inevitable presidential commission played out its part, but very little radioactivity was released, and no one was physically hurt. The psychological damage suffered by the population of the area was, however, no less real than physical damage.

The nuclear safety community went through a minipanic, and some anti-nuclear groups were inexcusably ecstatic about the accident. The Nuclear Regulatory Commission issued two hundred requirements for change, some of which have still not been accomplished, and a couple of hundred changes in response to a simple accident is a panic response. In the safety business stability is a virtue, and too much change, especially without analysis, is a bad idea. None, really none, of these mandated changes were analyzed to determine their net virtue. It was a rulemaking frenzy.

The accident itself was simple. At about 4:00 a.m. (accidents tend to happen in the wee hours), with the reactor operating at nearly full power, the reactor feedwater flow was interrupted. This automatically started a sequence of events (including a SCRAM) designed to keep the core cooled. Some pumps were tripped (shut down) and others started, including the two pumps of the auxiliary feedwater system. Unfortunately, the valves between these pumps and the reactor were closed, and it took nearly ten minutes to sort that out. They are never supposed to be closed, and no one has yet admitted to leaving them that way. Though not proved, maintenance, as in the Lockheed-1011 O-rings from the last chapter, was the likely culprit.

With the valves finally open, and the pumps operating, the incident should have been over. (Loss of feedwater is a frequent event even in coal-fired plants.) Sadly, a valve that had opened automatically during the initial pressure surge never closed properly, and no one knew it until too late. It was a tragicomedy, in which the not-quite-closed relief valve leaked steadily, the reactor bled to death, pumps were turned off in conformity with erroneous emergency procedures, the operators ignored clear signals in the instrumentation about what was happening (the offending valve had always leaked a little bit, so evidence of leakage was not taken seriously), and the full gravity of the situation was slow to sink in. Control was reestablished after about four hours, but the core damage had long since occurred. The damage sequence could have been reversed at any time within the first two hours, and few of us would then have known where to find Three Mile Island on a map.

The presidential commission (the Kemeny Commission) got things pretty straight. The accident began mechanically, was helped along by multiple human errors, and was an inevitable consequence of complacency. An industry that had never known an accident had come to believe it was impossible. Sound familiar?

Several years earlier, in 1975, a pioneering probabilistic risk assessment of nuclear reactors had been completed (the Rasmussen study). Despite other problems, it correctly concluded that the most likely causes of a core-melt accident would be transients (the initiating loss of feedwater was a transient), small LOCA (the leaking relief valve constituted a small loss-of-coolant accident), and human error (there was plenty of that). The accident, in one form or another, was predictable and inevitable. But neither the industry nor the regulatory forces reacted appropriately—they were delighted with the study's main conclusion that reactor accidents were very improbable. They were also under siege from the anti-nuclear forces, who liked to emphasize large LOCA. There was plenty of blame to go around.

There seems to be no therapy or prophylaxis for complacency, nor for the ills of a mature technology, except the sobering experience of frequent non-catastrophic accidents. If a probabilistic study predicts an accident probability that is not zero, and no responsible study will do otherwise, that means that the accident will happen—the only useful question is when. Failure to appreciate that can be damaging to the nation's health and well-being.

The flood of new regulations produced by the accident, however ill considered many of them were, has had a generally positive effect. Certainly the industry is more tightly regulated than it was in the past, which may not be all to the good, and somewhat more realistic, which is. Probably the most salutary effects of the accident are an increased emphasis on operator quality and performance, and recognition by the industry that an accident holds the seeds of economic disaster. Industry-sponsored groups have been set up to exchange safety information and, at least as important, to work toward a common standard of safety across the land. While it isn't possible to make everyone above average, it is possible to improve the average and to help the laggards.

One further observation, dear to the author's heart. This was the first major accident in the commercial nuclear industry (the first in the government nuclear facilities in this country is, at this writing, yet to come), and it would have rated a full NTSB-type treatment from impartial experts, to squeeze all possible lessons from it as dispassionately as possible. An institutional structure for doing that sort of thing had been proposed two years before the accident, but was of course successfully resisted by the NRC. (One of the arguments given at the time was that such an organization would have nothing to do, because there are no nuclear accidents.) Many events later, both small and large, there is still no such independent review body, which means that there exists no one with the responsibility to assess how well the regulatory organization, the NRC, is doing its job. The anti-nuclear people have no credibility, since their avowed objective is not to improve safety but to shut down nuclear power, and the NRC can hardly be relied upon to police itself. And Congress is no help.

So whenever there is a disaster that commands public attention, the President will appoint a special presidential commission to review it. This will always result in severe criticism for the operating organization, criticism that will fade with time after the commission is disbanded. Complacency will then be allowed to return, and business-as-usual restored. Self-criticism doesn't come easily to humans. After Three Mile Island, President Carter tried to maintain some level of surveillance by appointing a President's Nuclear Safety Oversight Committee, but President Reagan allowed it to die quietly. The most plausible explanation for the uncoordinated complexity of the nuclear regulatory system is that, deep down in their hearts, many participants still don't believe that a major damaging accident is possible. They are wrong.

Chernobyl

The nuclear accident in April of 1986 provided for most of us the first inkling that there was such a place as Chernobyl. By far the largest nuclear accident ever, it was the first to result in a large radioactive release. Much of the reactor core was blown into the sky, and the radiation was detectable over the entire western Soviet Union, Scandinavia, and Eastern Europe. The accident was first detected by the Swedes, who feared it was a problem with one of their own reactors. Because the damage to the Chernobyl reactor was so complete, the event can serve as a kind of upper limit for what can happen in a nuclear accident. It is hard to imagine worse.

The accident began (of course) in the pre-dawn hours, at Unit Four of a four-reactor complex near the large city of Kiev, in the Soviet Ukraine. This reactor, unlike Three Mile Island, was not running at full power; it was hardly running at all. The operators were performing a test to show that the reactor could withstand a certain kind of hypothetical accident; the point of the test hardly matters.

As part of the test, the reactor was supposed to enter an operational regime that is forbidden in normal operation, and which would result in a SCRAM if it ever happened in real life. To prevent a SCRAM during the test the operators therefore disabled many of the safety devices, so that when the reactor was in real trouble it had no automatic means of escape. The accident needn't have occurred, but the operators were apparently confident that this type of reactor was so safe that accidents were, in effect, impossible. They were complacent—that ought to sound familiar by now.

To understand this accident we need to know a bit more about reactors. The reason they work at all is that a fissioning nucleus releases particles called neutrons, which are themselves able to induce fission in other fissile nuclei. Thus it is possible to set up a chain reaction in which a nucleus undergoes fission, releases neutrons, which then induce fission in another nucleus, which then releases neutrons, and so on. Since the fissioning nucleus releases several neutrons, and it only takes one to maintain the reaction, there are plenty of neutrons to go around. All reactors therefore have some provision for absorbing the surplus neutrons, through control rods, neutron-absorbing materials, or by just letting them escape from the core. The control rods absorb neutrons. All it takes to SCRAM a reactor is to insert lots of neutron-absorbing material fast. So far, so good, but you have been shamefully deceived.

What has been described wouldn't work at all, because the reactions are so fast (a very tiny fraction of a second) that if there were a surplus of neutrons around for just an instant, without enough absorber, the chain reaction would grow out of control long before anything could possibly be done to stem the tide. That is how an atomic bomb works, and it's hard to stop an atomic bomb in mid-explosion. But nature is kinder than that. A small fraction, about 1 percent, of the neutrons from a fission are so-called delayed neutrons, taking a full second or so to be born, and that one second of grace makes it possible to control the reactor. Some neutrons take more time, others less, but a second is a reasonable average.

Reactors operate in a regime in which the effective number of instantly released neutrons per fission (called prompt neutrons) is very slightly less than one, and the difference between that and the needed number of exactly one is made up of delayed neutrons; they can be controlled. It is a delicate balance, and is the basis for the controllability of all reactors.

But there is an important cautionary note here. If ever a control rod is pulled out too far, so that a chain reaction can be achieved with the prompt neutrons alone, disaster will ensue; the reactor will have gone "prompt critical." That happened at Chernobyl. The details are irrelevant— it is another tragicomedy, involving a fundamental design flaw and lots of human error—but it is truly unthinkable that a reactor operating crew could let it happen.

After the accident, which blew off the top of the reactor and dispersed radiation everywhere, the Soviets reacted heroically and well. Those who were overexposed to radiation (in the end, thirty-one people died) were mainly firefighters, who fought a raging fire in a deadly radiation environment, and were true heroes. One can only admire the Soviet response, and their later willingness to share the accident information with the international community, which they did in detail that summer. It is notable that the Soviet report to the International Atomic Energy Agency begins by saying that this accident will not turn them away from nuclear power, because nuclear electricity generation is least harmful to the environment. The reactor at Chernobyl is now entombed in concrete, but the other three reactors on the site are operating.

As of early 1990, the two additional reactors under construction at the site (which were of the same type) have been cancelled, and the long-term future of the three operating reactors is unclear. There has been a major reevaluation within the Soviet Union of the quality of their nuclear program, the inevitable reorganization has occurred, and many old-type reactors that were either planned or under construction (including all reactors of the Chernobyl type) have been cancelled. In addition, *Glaznost* has unleashed substantial anti-nuclear sentiment, with unpredictable consequences for the program.

As a side note, the Chernobyl reactor was of an unusual design that uses graphite to slow down (moderate) the neutrons—slow neutrons produce fission better than fast ones—and there is one large reactor in the United States that also uses a graphite moderator. That old reactor is owned by the Department of Energy, and is located at Hanford, Washington. It was decided in the end to shut it down permanently, not because an event like the Chernobyl accident was likely, but because it was not up to modern safety standards. The reactor used to produce plutonium for the nuclear weapons program. Few if any of the reactors owned by the Department of Energy are up to modern standards, and our country has a problem.

EMERGENCY RESPONSE

If there is a Chernobyl-like nuclear accident, what is the real threat to the public? The risk comes from the release of radioactive materials normally confined in the core of the reactor, which is in turn enclosed by the familiar dome-shaped containment building. (At Chernobyl there was no containment building—some Soviet reactors have one, and some don't.) The radiation is lethal stuff, not that a whiff of it will drop you in your tracks, but for the reasons mentioned at the beginning of the chapter—radiation can cause cancer.

The most important emergency-management issue in the event of a major nuclear accident is clear: to minimize the exposure of the public to radiation. In the worst cases one would also need to protect the food and water supplies, and deal with land recovery and cleanup. Smaller exposures to radiation do less damage, and there is a point below which the damage, if any, is beneath notice.

The first axiom about the management of a nuclear accident (and most other emergencies) is that we shouldn't worst-case ourselves out of a solution; we shouldn't base all our planning on the worst things we can imagine. They are the least likely, and planning for the worst can leave us unprepared for reality. Imagine a fire department spending all its training, and purchasing all its equipment, to fight a major "towering inferno" fire in a hundred-story skyscraper. The firemen would go through life waiting, and would have trouble dealing with the wastebasket and garage fires that are everyday events for a typical fire department.

Unfortunately, nuclear accidents are so feared that it has been impossible for emergency planners to avoid worst-case scenarios; nearly all planning is for the big event. This vicious circle leads to what a control-systems engineer might call an instability—the fact that one plans mainly for the big accident makes it seem more plausible, which in turn means that one had better plan for it. As a consequence, the public never hears that there is any possible end for a nuclear accident, short of monumental destruction. The planners soon come to believe it themselves.

There are three ways to reduce radiation exposure from a nuclear accident: to wait, to run, or to hide. All radioactive material does decay (i.e., become harmless) in time, some in seconds or less, some in minutes, some in days, and some in millions or billions of years. Contrary to common belief, the short-lived materials are the most radioactive, while the long-lived ones are hardly radioactive at all. It makes sense to take shelter long enough to wait out the short ones. In addition, some of the radioactive materials—all of them at the beginning of an accident—are airborne, and prevailing winds will disperse them. The two reasons to wait are radioactive decay and winds.

Ordinary building material provides reasonable pro-

tection against radioactivity. Getting indoors, hopefully
at home and preferably in the basement if there is one, is
a reasonable procedure. One should, of course, turn on the
radio or television set, to get information and advice from
the local emergency experts, *not* from the evening news
programs. The latter did very badly at Three Mile Island,
where they fomented panic in an accident that, in the end,
did most of its damage to the utility stockholders.

But you can't wait indefinitely. The most damaging
radioactive materials have half-lives (the time it takes to
decay to half strength) of about thirty years, and something has to be done long before then. When the waiting
and hiding are over, there will be a proper time to get out
of the neighborhood, to cut and run. The time may be
hours or days or never, depending on the details of the
accident, and should be determined by the experts who
will by now have been mobilized. This sequence of mitigative actions, sheltering followed by relocation, is optimal
for nearly all major nuclear accidents, and is close to the
Soviet response at Chernobyl. The Soviets were wrongly
criticized by anti-nuclear groups in the United States for
not having ordered an immediate evacuation. They handled their unprecedented emergency well, heroically in an
impressive number of individual cases, and declined to be
panicked into premature evacuation. When it needed to
be done, they did it. Perhaps they were a bit slow, but
not by much.

How does this compare with our planning for such a
contingency? Well, we emphasize early evacuation. We
pay lip service to the wait-and-hide strategies, but our
emphasis is on cut and run. To get a license to operate a
nuclear power plant, the utility involved must demonstrate
that it is possible to evacuate all the people within either

ten or fifty miles of the accident (depending on details),
regardless of the wind direction. This obviously requires
the cooperation and participation of the local and state
authorities, who are in charge of all emergency response,
and the rules require frequent combined drills. Some local
and state officials are opposed to nuclear power—that's
good politics these days—and have learned that they can
block the licensing of a plant by simply refusing to participate in emergency evacuation drills. Governor Dukakis of
Massachusetts has used such a maneuver to obstruct the
licensing of a plant in New Hampshire, a neighboring state.
There is, at this time, a similar situation on Long Island,
in New York, with Governor Cuomo having switched to a
popular anti-nuclear stance.

Most of the above is relevant to more than nuclear
emergencies. Planning for most emergencies needs to be
thoughtful, with strategies that contain enough flexibility
to deal with a broad spectrum of contingencies. It is a mistake to plan in too great detail, because such planning inevitably emphasizes the largest possible disaster, and small
contingencies are not covered by planning for the worst.
Besides, one can't really predict the sequence of events. It
is important to distinguish between strategies, which are
broad and flexible outlines of an effective course of action,
and tactics, which tend to be detailed blueprints. Only
bad architects let the latter dominate the former. Former
President Eisenhower once said, "Plans are worthless, but
planning is everything."

Without effective leadership no tactics will work; there
must be a prior understanding of who is in charge. It is
said that in the military the issue is decided by counting
the number of stars on an officer's shoulder; it is sufficient
to be able to count to five. It matters less who stands

under the stars than that everyone agree on the same person. A friend of the author, a former military man, says that the quality of command is never important if you are winning, but that it is crucial if you are losing. For a nuclear accident, experience shows that the highest available government official will take charge—Mr. Gorbachev did so for Chernobyl, though he had the good sense to send a competent scientist to the scene—and that that official will operate out of his instincts. Since he will have had little or no training for this sort of thing, his performance will depend largely on his choice of advisors.

Prior arrangements for leadership are essential even for less well-publicized emergencies. In California and perhaps elsewhere, where small fires sometimes grow across jurisdictional lines, fire departments understand that the fire chief in the area in which the fire originally started remains in charge, with other fire departments lending him support, no matter how far the fire spreads. Obviously reasonable people can negotiate a transfer of authority if needed, but the principle of maintaining an unambiguous command structure is sound. The relevant sentence from the California Mutual Aid Plan (originally signed by then Governor Earl Warren in 1950) is, "The responsible local official in whose jurisdiction an incident requiring mutual aid has occurred shall remain in charge at such incident including the direction of such personnel and equipment provided him through the operation of such mutual aid plans." No ambiguity there. If it is clear who is in command, the necessary flexible strategy has been thought through in advance, and that person has both adequate provisions for two-way communication with his forces and good advice and intelligence (both kinds), giant steps will have been taken toward management of the emergency. In these matters, nuclear emergencies are orphans.

HIGH-LEVEL WASTE

Fission products are radioactive, and the used fuel rods from a reactor are loaded with them. In a typical reactor, about a third of the fuel rods are changed each year, a complete turnover every three years or so, and something must be done with the used rods. They are far too radioactive to handle, and are normally first stored under water at the reactor site. The radioactivity will decrease gradually, and the water will provide the cooling for a long time. The problem is what to do with the rods after they have "cooled" to a reasonable level of radioactivity—that takes a decade or two.

The used rods are not exactly waste. They begin their lives as a mixture of isotopes of uranium, of which the most important have weights 238 and 235. U-238 is by far the most common, but is not particularly fissile, while U-235 is fissile, but comprises only about 0.7 percent of natural uranium. For American reactors it is customary to increase the percentage of U-235 to about 3 percent or 4 percent, because that makes it possible to use ordinary water as moderator. Ordinary water is cheaper and more plentiful than heavy water, used by the Canadians in their commercial reactors and by us in some government-owned reactors. The remainder of the uranium is U-238, which produces plutonium when struck by a neutron. (The process is a bit more complicated, but that is the usual outcome.) Since plutonium is fissile, it would seem reasonable to reuse it, but plutonium is also useful for nuclear weapons. President Carter therefore decided long ago that the United States would set a good example for the world by not reprocessing spent fuel to recover the plutonium, so we don't. This leaves the plutonium in the spent fuel, and makes the disposal problem harder. And as a good example, our gesture was a bust.

(Despite what may have been inferred from the above, this author also advocates an energetic non-proliferation policy, and is not averse to strong measures in support of worthy objectives. However, he has little faith in gestures of goodwill as means of deterring nuclear deception. There are countries that lie and cheat to hide their surreptitious nuclear weapons programs, and it will require more than gestures of goodwill to discourage such countries.)

The waste decays, and must be kept out of the human environment while it still has the potential to do harm. There is no easy definition of just when that point comes, but in a few hundred years the waste has decayed to about the same degree of radioactivity as the original ore from which it was produced. After that, if the waste has been buried, radioactivity will have, on balance, been removed from the earth, and the canisters can crumble as they wish. Of course they won't.

There is an important point here, key to understanding high-level waste problems. All radioactive materials have a natural half-life, a time in which half the material decays into something else. That something else may or may not be radioactive—if so, it has its own half-life. Or there may be two or more decay products, any or all of which may be radioactive. If the half-life is, for example, one day, then half the material will be left after a day, one fourth after two days, one-eighth after three days, and so on. High-level waste from reactors is a mixture of over a hundred different materials, each with its own half-life, so the rate at which it decays is a complex function of all these. After the short-lived materials have decayed, what remains will be the longer lasting of the original ingredients.

A nucleus can only emit radiation when it decays, and that can take a while if it has a long half-life. So the specter of high-level waste remaining extremely radioactive for hundreds of thousands of years is just fiction. Long life and high radioactivity are antithetical. That is why the waste is not very threatening after a few hundred years—the hot short-lived ingredients have long since decayed, and all that is left is relatively benign. Not entirely benign, but relatively benign.

The mixture in the waste divides very roughly into two classes, those nuclei with half-lives less than a few hundred years, and those with very long half-lives, like millions of years. Therefore, it would seem that if the waste were kept isolated for a few hundred years, only the relatively gentle stuff would remain, and all would be well. This is what would have happened if the original plan to remove the plutonium for reprocessing had not been discarded. But there are two major isotopes of plutonium in the waste, one whose half-life is 6,000 years, and one with 24,000 years, longer than the short stuff, and more radioactive than the long. That makes things harder.

The radioactivity of typical waste decays by about a factor of 100 in the first year after the reactor is shut down, an additional factor of 5 in the next ten years, and another factor of 10 in the next hundred years. After that the decay depends on the particular constitution of the material, but it is steady, and no way is known to either speed it up or slow it down.

The disposal plan (as of late 1989) is to pack the waste with its plutonium untouched, and bury it in deep caverns in a mountain in Nevada. Nobody wants it, and it will

have to be jammed down Nevada's throat; that may not
be possible. Standards for the repository have been established by EPA, and require that it remain intact for ten
thousand years, by which time the radioactive materials
will be relatively innocuous. EPA requires that the waste
be stored in such a way that future people, presumed to
be ignorant savages, will not be able to hurt themselves if
they accidentally dig the stuff up. Apart from the patent
arrogance of that presumption, future people will probably
welcome the discovery of these canisters; they will contain
materials that will surely be in short supply by then.

The arrogance deserves emphasis. We assume that we
know much more than the people of the past, and it is
even true, certainly as far as science and technology are
concerned. To assume that we are also more competent
than the people of the future means that we have selected
ourselves as the highest manifestation of the human race,
the peak of human development for all time. An engaging
thought, just a bit pompous.

Of course the waste is not to be dumped into a hole in
the ground, and then covered over with a tarpaulin. The
current intent is to modify it into a particularly hardy
form, molded into solid glass, impervious to any ground
water that may penetrate the repository, and then encase
it in further impervious and durable canisters. Finally, it is
to be buried in a deep, stable, geologic repository, there to
await disturbance by either geology or future generations.

The amount of high-level waste produced by a single
large reactor in the course of a year's operation is a few
cubic yards, the capacity of a light pickup truck. Of course
it is more dense than the usual pickup load, and might
bend the frame, but these are not mountains of material.

It is possible to estimate the risk if the material is

buried as planned. It turns out to be ridiculously low, at least a factor of a million lower than any other risk discussed in this book. The risk is as negligible as it is possible to imagine, yet the clamor about the subject has paralyzed the decision-making authorities, and there is still no consensus solution. It is embarrassingly easy to solve the technical problems, yet impossible to solve the political ones. Other countries seem to find the going easier, possibly because we pay a price for a government so utterly barren of people with any technical education. That wasn't always true—Benjamin Franklin was an accomplished scientist. High-level nuclear waste disposal is a non-risk.

Here is a true story for those who are most worried that leakage from the repository will find its way into the biosphere after ten thousand years. Though we are the first ordinary humans to have discovered how to build nuclear reactors, nature did it long ago. About two billion years ago, in what is now the African country of Gabon, conditions were just right for some natural uranium deposits to undergo nuclear fission, and to actually start and maintain a chain reaction. No one really knows how long it lasted, perhaps a few hundred thousand years. This is fascinating in itself, but the relevance to high-level waste disposal is that nature didn't have all these fancy canisters and caves in which to store its nuclear waste; the waste was just left where it was, and where it still is after two billion years. It would seem that with carefully selected geology, modern technology, well-built containers, human ingenuity, and deep burial, it should be possible to do a millionth as well. Our planned repository has to last about as long as some of the Roman ruins or possibly the Egyptian pyramids, and if our engineers can't match that technology we are in deep trouble. Even EPA, promulgator of the current safety standards for the repository, predicts that "it will cause no more than 1,000 deaths over the entire 10,000 year period." Note that "no more than" means an upper limit, conservatively estimated, and the most likely number is much smaller. Even the upper bound corresponds to one death per decade, in a country in which two million people now die each year. It is a phoney issue, and it would be in the national interest to get on to more important matters.

Radon

When we listed the "normal" exposures of Americans to ionizing radiation, radon and its progeny were explicitly overlooked. We must now make up for that omission, and it was a large omission indeed. Strictly speaking, the subject of radon doesn't belong in a book about technological risk—radon is natural, just as cosmic rays are natural. Cosmic rays from the sky, radon from the ground—how neatly complementary.

More than a century ago it was recognized that underground miners, working in the Erz Mountain area near the current border between East Germany and Czechoslovakia, were especially prone to lung cancer. It had even been known a couple of centuries earlier that the miners tended to develop a serious respiratory disease, which was known as "miners' disease." These were mysteries, and only relatively recently was it finally learned that the problems came from breathing radon gas and its products, emitted by the radioactive rocks. All rocks are somewhat radioactive—that is the earth we inhabit. There are now several other cases involving different groups of underground miners, including American uranium miners working on the Colorado plateau. Radon gas and its products predispose to lung cancer.

Radon is the product of a long chain of radioactive events beginning with the parent uranium isotope U-238. Uranium is ubiquitous in the earth's crust, in concentrations averaging around one to a few parts per million (by weight). U-238 has a half-life of 4.5 billion years, approximately the age of the earth (which is why there is still some around), and decays through five steps that lead to radium, which in turn decays to create radon. Radon itself has a half-life of 3.82 days, and, most important, decays through the release of a high-energy helium nucleus (called an alpha particle) into a variety of relatively short-lived progeny, which in turn decay, also through alpha particle emission, to do most of the damage to human lungs. It is not a simple process.

Radon is a noble gas, not because it has royal ancestors, but because it is incapable of chemical bonding, as are helium, argon, neon, etc. Noble gases are also called inert gases, again without any social implications. Therefore the radon we inhale is just as quickly exhaled, and is unlikely to decay during its short stay in our lungs. If it doesn't decay, it does no harm. But air containing radon always contains a certain concentration of its decay products (called progeny). The progeny are not inert gases (lead, polonium, and bismuth are the main culprits), so they can and do stick to dust particles and to the lining of the lungs. There they are free to wait around and decay at their leisure, releasing the alpha particles which are, in the end, responsible for the damage. It is the progeny that are most harmful, not the radon.

How much radon is there in normal air, and how bad is it? The primordial uranium is present in all soil, so radium (a decay product of uranium, and the immediate parent of radon) is also present, at lower concentrations. Since the half-life of radium (about 1,600 years) is much shorter than that of the uranium, there is in effect one radium decay into radon for every uranium decay that ultimately produces radium. (It is as if a person is a spendthrift living on a meager allowance. He spends the radium by turning it into radon, as fast as he can get it from the uranium, just as he might spend his income as fast as the allowance checks arrive. The rate of expenditure is determined by the rate of arrival. A real spendthrift might be able to cover a shortage with a loan, but nature doesn't permit that.) So a measure of the amount of radon produced is the amount of uranium that decays. Since radon is an inert gas, some of it seeps out of the top layers of soil during its ordained few days on earth, and drifts in the atmosphere, awaiting its fate. That fate is of course decay.

The unit for amounts of radioactivity is named the curie, after Pierre Curie. Curie won the Nobel Prize, with his equally distinguished wife Marie, for the discovery of radium. (He was killed in Paris in 1906 when run down by a horse-cart.) The curie, abbreviated Ci, was originally defined to approximate the radioactivity in a gram of radium. More useful is the picocurie (pCi), defined as a millionth of a millionth of a curie, a decay every thirty seconds or so. The average radioactivity of the air over typical outdoor soil is about a tenth of a picocurie per liter, $0.1\,\mathrm{pCi}/\ell$, and a liter is about a quart. (Some day the metric system will come to stay.)

Like the cosmic rays, radon is everywhere, effectively inescapable, and our outdoors exposure to it is approximately the same as our irradiation from cosmic rays. The special problem with radon and its progeny is that they accumulate in buildings (and mines), where the air stagnates. Outdoors they are swept into the upper atmosphere by the prevailing breezes as fast as they are released from the soil, so they are heavily diluted. Houses, however, rest on the ground in one way or another, and are more or less airtight. Obviously the concentrations in the indoor air depend on the kind of soil under and around houses, the pathways from that soil into the house, and the degree of ventilation. When our mothers told us that fresh air was good for us, they were right, even if their hidden motive was simply to get us out of the house. Ventilation and construction can affect both the concentration of radon in a house and the amount of the accumulated progeny. Though the effects can go both ways, the concentrations indoors are typically greater than outside.

To make this point a bit more explicitly, the outside air has an average concentration of radon and its decay products that is determined by the kind of soil and by the average winds. Since radon has a half-life of a few days, there is plenty of time for it to get thoroughly mixed with the air after it has seeped from the ground, and the concentration in the air depends simply on the rate at which the ground supplies it, and the rate at which it is swept into the upper atmosphere. If it is being swept away rapidly, there won't be much around, and vice versa. The progeny have somewhat different careers. They are chemically active, and are able to stick to the soil while the radon is seeping through it. Therefore the radon that emerges from the ground has had some of the progeny swept from it. Once free of the soil the radon then sets out (by decay) to replenish the supply of progeny, which it does over the next half hour or so. The reason for the half hour is that the progeny half-lives range from a few minutes to a half hour. House interiors can accumulate large concentrations of radon progeny by keeping the radon in place long enough to decay, especially if the ventilation is poor. And the progeny stick to the lungs long enough to themselves decay.

It is hard to directly compare radon-induced irradiation with that from other natural sources. The radon progeny do their damage in the lungs and other parts of the respiratory system, while cosmic rays affect the whole body, and medical and dental x-rays affect certain parts of the body more than others. It is usual to find a common ground, however imperfect, by expressing all these exposures in terms of equivalent "whole-body dosage," so they can be compared. With that convention, one picocurie per liter from radon is approximately equivalent to 200 millirem per year from cosmic rays, so that the effective dose from radon for the average American is as much as from all other sources combined. That is impressive, and is the reason that radon has become a major public health concern.

We don't know if this is damaging to the health—the uranium miners, who are the best sources of information about the damage potential of radon, were exposed to far more. It's the familiar problem of the effects of low levels of radiation—damage so small that it can't be detected, and no direct evidence that there is any damage at all, with prudence dictating that we assume there is.

In this case, however, there are two factors that make the risk more real. The first is the large variation of concentration among houses. Very limited surveys have shown big differences even from house to house within a given area. Some houses yield a thousand times more radon exposure than others, and the worst houses show concentrations approaching those experienced by the uranium miners. There is no potential for widespread disaster or

for people dropping in the streets, but there is some risk
for some of us.

EPA recommends caution and attention to the problem if a preliminary screening test in the closed house reveals radon activity greater than $4\,pCi/\ell$, immediate action
if the number exceeds 200, and inaction if the number is
less than 4. These are not unreasonable recommendations,
though the risk, if any, is exceedingly small at the lower
levels.

The second special factor involves smoking. Lung cancer is still epidemic in the United States—nearly 150,000
new cases were diagnosed in 1986, and the median time
between diagnosis and death is less than a year. It is a
dreadful disease. The most common cause of lung cancer is
of course smoking, responsible for nearly 90 percent of the
deaths, according to the Surgeon General. The lung and
the rest of the respiratory tract are precisely the locations
also attacked by radon and its progeny. While they alone
make only a small addition to the risk of lung cancer, the
fact that smoking and radon attack the same areas opens
the possibility that their effects are additive or mutually
supportive. Is a person's susceptibility to radon-induced
cancer enhanced if he or she is also a smoker, beyond the
simple sum of the two effects?

The best way to answer such a question is to study
comparative mortality statistics among smokers and non-
smokers who are or are not exposed to radon daughters.
There are thus four groups to look at. Since smoking is by
far the larger of the two effects, it is best to concentrate on
the population whose exposure to radon is best known—
the miners—and try to unscramble their smoking histories.
This is difficult, especially among those who are deceased,
but it has been done for the small sample that is avail-

able. At this writing, the most authoritative analysis of the data is that published by the National Research Council in 1988, which begins by acknowledging that the data are really inadequate to the job (and therefore, of course, that more research is needed). It then concludes that the existing evidence suggests that the effects are more than additive, which implies that smokers have a special reason to keep away from radon-filled houses. They could, of course, accept the house and drop the smoking.

16
Fossil Fuels

FAMILY CHARACTERISTICS

Chapter 13 described part of the trauma inflicted on the country by the 1973–74 Arab oil embargo. Temporarily (and, alas, only temporarily) the embargo shocked us out of a comfortable but false sense of security about our energy supply.

Energy is one of those words whose everyday meaning has little to do with its technical meaning. Scientists define energy as the ability or potential to do work, but work is another such word. Beginning physics students are often shocked (and a bit incredulous) to be told that if they hold an anvil over their heads for an hour, they will have done no work. Work involves applying a force (another such word) over a distance, a word which, believe it or not, means exactly what it says. That should clarify everything.

That confusing little exercise was meant to illustrate the pitfalls of nomenclature that separate what C. P. Snow called the two cultures. They are often not even speaking the same language. Of course the scientists could have given these precise concepts names that weren't already used for other purposes, like wrokitude or forfinence, in which case there would have been no problem of misunderstanding, or even understanding.

Energy is best thought of in terms of its sources, which are, for the United States: fossil fuels (coal, oil, and natural gas), the combustible remnants of ancient living objects like trees and vegetation; hydropower, solar energy captured by letting water evaporate in sunlight, to be deposited at higher altitudes; nuclear energy, the energy stored in some of the original constituents of the earth; and a few smaller contributors. For most of human history energy was obtained through animal power (power is the rate of use of energy), and the most commonly used animal was man himself. We use what is called biomass conversion; the foods we eat (chemical energy) are converted to other forms of energy, which is then used to do work. We are motors fueled with food, and most of our expendable energy was used in the past to keep ourselves fueled.

The domestication of stronger animals to do our work was a major step forward in human well-being. An ox can do nearly ten times as much work in a day as a man, and a horse even more. The term "horsepower" was invented to describe the average rate at which a good horse can do work for a day, so mechanical engines could be directly compared to the standard horse. (Actually, real horses fall short of a horsepower, over a full day's work. James Watt, who invented the term, must have thought his horses weren't doing their level best.) Nowadays even small automobiles and tractors can do a hundred times as much work as a horse, so the days in which we were limited to human muscle are long gone. It is easy to underestimate the importance of this emancipation. The *average* rate at which we consume energy in the United States is equivalent to having about fifteen horses working for each of us, full time, and working hard. That would require a stable of about fifty horses for each of us—even horses have to rest.

The most important single fact about energy is that it can't be created out of the whole cloth. All we can do is change the less useful forms of energy into more useful ones, collecting the original supply from nature. That's all the sun does, burning the constituents it inherited when it was formed. Fortunately, the sun was supplied with enough energy to last quite a few billion years, so there is no immediate cause for alarm. This deep fact, the conservation of energy, was not known to the ancients, and its full implications have only become clear during the last century and a half, first with the recognition in the mid-nineteenth century that heat is a form of energy, and continuing through the realization by Einstein at the beginning of this century that mass could be turned into energy, and vice versa. We now know many forms of energy, and much more science than ever before, but there is no sign of any weakness in the energy conservation law. We use what we inherited, and can do no more.

All sources and uses of energy can be lumped into what is called the energy budget, income *vs.* outgo, and the United States has been running a deficit for many years. The deficit has long been covered by importing oil from other countries, sending them money in return, and that contributes mightily to the current trade deficit. (About a third of the trade deficit, nearly $50 billion a year, is now spent on imported oil, and that can only increase.) For a short time after the 1973–74 embargo we cut imports through conservation, but are now back where we were, at a higher price. We will eventually pay the piper for this prodigality. By 1988 we imported nearly half our total supply of petroleum, and the fraction is growing. The biggest suppliers are Saudi Arabia, Canada, Venezuela, and Mexico, and the reader can make his own judgment about the political stability of the supply.

When energy is converted from the form in which it is found to a more useful form, one of the immutable laws of nature exacts a cost. Energy is not lost—that is the first law of thermodynamics—but the second law decrees that it can't all be converted. The fraction of useful work extracted from a source of energy is called the efficiency of the process, and is always less than 100 percent. Not only do you get less out than you put in, it is usually much less. History is littered with the detritus of inventors who have given us the designs of "perpetual motion" machines, violating either the first or the second laws of thermodynamics, or sometimes both. They are all frauds, many fooling even themselves. Like cancer quacks and other pseudo-physicians, they continue to thrive, and only do harm when people believe them.

Heat (another such word, not the same as temperature) is a form of energy, and can be made to do work—witness a steam locomotive or even a car—with the losses in inefficient conversion of energy usually turning up in discarded heat. The human body is about 25 percent efficient in converting food energy to useful work, 75 percent of the original supply going into waste heat. That is why we get "overheated" if we exercise vigorously—our bodies are working to dispose of that waste heat. They do so by raising the temperature of our skin, and eventually by sweating.

So there are inevitable losses in energy conversion, which we accept as the price of changing the energy into a useful form. More than a third of our total energy supply goes into the production of electricity, most often at 30–40 percent efficiency, though hydroelectric conversion is very efficient. (Most of the good hydropower sources are already in use, and exploitation of the remainder would cause unacceptable environmental disruption.) Gasoline and diesel engines are about 20–25 percent efficient, solar photovoltaic cells (much oversold, but improving) about 5–20 percent efficient, electric motors (once the electricity has been made from something else) about 90 percent efficient, and so on.

Most of our energy comes from burning fossil fuels (nearly 90 percent in 1987), and the supply is running out. The fuel with the longest potential life is coal, which might last for a few hundred years at the current rate of depletion—estimates differ. There is no known solution to the long-term problem of energy supply, especially with a growing world population, and with the Third World aspiring to the standard of living of the industrialized world. Here too the many purveyors of easy but illusory solutions are no help.

There is one amusing fact about the energy supply, pointed out by Richard Wilson, which may help to maintain perspective. All coal contains both uranium and thorium in trace amounts, typically a few parts per million. If it were possible to collect the tiny amounts of uranium and thorium from a ton of coal, and burn them in a nuclear reactor, that would provide ten times the energy available from burning the coal directly. Of course it isn't techni-

cally or economically sensible to do this, but it illustrates
how much energy there is in nuclear fuel. An atom of
uranium-235 has a hundred million times as much available energy as a carbon atom in coal. Pound for pound,
as we have said, the ratio is nearer to ten million.

Fossil fuels produce energy mainly by burning carbon
and hydrogen, as do the foods we eat. Coal is mostly
carbon, while petroleum has carbon and hydrogen, and
natural gas has more hydrogen than carbon. To burn such
materials we combine them chemically with the oxygen in
the air, releasing heat. That is then either converted to a
more useful form of energy or just used for warmth, while
dumping its chemical wastes into the atmosphere. The
waste from burning carbon is carbon dioxide, CO_2, while
that from burning hydrogen is ordinary water, H_2O. These
are the same waste gases we exhale as we burn our own
fuels, whether chocolate bars or rib roasts. We have no
choice but to dump them.

There are other incombustible components of coal and
oil, and these either form the ashes that are left over or are
themselves dumped directly into the air. In modern coal
plants scrubbers are used to reduce the releases of these
problem residues, and the scrubbers can be pretty good.
They aren't perfect, and the wastes that are turned loose
cause some of the environmental and health hazards that
are the price of burning fossil fuels. The biggest problem,
and it is truly apocalyptic, is caused by the carbon dioxide.
That is for later in this chapter.

Many years ago Otto von Frisch wrote a whimsical article describing an imaginary society whose members have
long lived with nuclear power as their only source of energy. They have known for some time of buried deposits
of some black material, which is used for coloring mat-

ter during tribal ceremonies. Finally they discover how to ignite this material and release energy. Of course they recognize immediately that the effluents, both products of combustion and contaminants, are dangerous, so it would be unthinkable to release them to the atmosphere, and the apparently insoluble problem of waste disposal keeps them from exploiting this otherwise promising source of energy. Some suggest crazy schemes, like collecting the waste gases in giant balloons, to be placed in orbit, but those are visionary solutions. There seems to be no good way, so they quite properly turn their backs on the discovery, and live happily ever after with a clean atmosphere and plenty of energy.

It is not so far-fetched. If we hadn't acquired the habit of burning fossil fuels and dumping the garbage into the atmosphere, it might not be possible to start now. But we are already hooked. An addiction is an addiction.

There is more to the balance and composition of the atmosphere than just the effects of burning fossil fuels, but they are dominant. The habit of dumping the waste products, just like all our dumping habits, dates from a time when the effect of humans on the planet was small, so that "natural" processes could easily absorb our debris. So effective was this that archæologists rejoice when they find any surviving relics or artifacts of old civilizations—most are long gone, and have left no trace at all. What has happened to turn this old habit into a current problem is that man has become a major component of the planet, with no visible end to his growth. As was said in the Introduction, the population growth problem will come home to roost (a bad metaphor) before the pollution problems make their own transition from merely damaging and offensive to intractable.

Acid Rain

Coal and petroleum don't consist of pure carbon and hydrocarbons, and even natural gas is impure. Their combustion products are therefore mixed with a variety of contaminants, mostly bad. Even if the fuels themselves were pure, the air they burn to release energy isn't pure oxygen—it contains about 78 percent nitrogen, which is chemically affected by the high combustion temperatures. More noxious effluents are formed at the high temperatures inside an internal combustion engine. About 40 percent of our oil goes into feeding cars and trucks, and half our coal into making electricity.

Among the products of combustion, sulfur dioxide, SO_2, formed from the sulfur contaminants in coal and petroleum, gets most of the blame for acid rain; most of our control efforts have been directed at SO_2.

What is the problem? The word acid usually conjures up an image of the vinegar (acetic acid mixed with water) that flavors our salads, or the more concentrated sulfuric acid solution that fills our automobile batteries. There is however a precise definition in terms of a measure with the unlikely name of pH. The opposite of an acid is an alkali, like lye, and the pH measures the extent to which a material is on one or the other side of the line. Pure water is considered neutral, and has a pH of 7.0, while acid solutions go down to 2 or 3 or even lower, and alkalis up to 10 or 12 or even higher. When we suffer from acidosis, our tummy pH is too low, and billions of dollars change hands every year as we ingest proprietary alkalis in an effort to "neutralize our stomach acids." The functioning of all living organisms, including humans, is sensitive to the pH of the environment, and depends on maintaining a stable inside level. The normal pH of human blood is 7.4, slightly on the alkaline side of neutral. Every gardener knows that rhododendrons thrive best in rather acid soil, with a pH between 4 and 5.5, while cabbage prefers a slightly alkaline soil, near 8. The pH scale is logarithmic; additive changes of pH lead to multiplicative changes in acidity. A change of pH from 5 to 4 means ten times as much acidity.

Though pure distilled water has a pH of 7.0, the rain that falls from the sky has never been pure. It has been distilled by evaporation from the surface, but its long journey before falling as rain gives it time to dissolve some of the components of the atmosphere. That changes its pH. Long before people started dumping their garbage into the atmosphere, the rain that fell was somewhat acid, and that is the stuff to which the earth became accustomed. The mechanism is simple: the air contains a certain amount of carbon dioxide from the decay of organic matter, and some of that dissolves. That turns normal rain into very weak soda water, or carbonic acid. Think of it as flat club soda. Typical rainfall might have a pH of about 5.6, on the acid side of neutral but acceptable to our flora and fauna.

In recent years certain parts of the world, including the northeastern United States and eastern Canada, have seen the pH of their rainfall head steadily downward, with severe damage to crops, fish, and trees. Even the famous German forests have suffered. It is our fault—we humans are feeding the atmosphere with enormous quantities of soluble gases, to be picked up by the rainwater. The main new culprits are SO_2, sulfur dioxide, and the oxides of nitrogen. Both are from fossil-fuel burning.

Some coals do have less sulfur than others, as do some oils. The low-sulfur fuels are obviously preferable, but also more expensive and remote from their points of intended use. Scrubbers for coal plants have already been mentioned. The best solution is to burn less—there is less of a problem in France, where 70 percent of the electricity is now nuclear.

If we insist on burning coal, the cost of doing it cleanly is high, and we are reluctant to pay it. Acid rain, however, shows no respect for national borders, and a nation unwilling to pay the price for domestic reasons may be forced to do so through its international relations. The penance is an inevitably higher cost of energy, whether through combustion technology, the use of cleaner fuels, or even the use of less energy. Any of those will force a reduction in living standards, which few nations will accept voluntarily.

So far our Congress has had the "courage" to pass a law, the Acid Precipitation Act of 1980, which mandates study and assessment of the problem, and we have even promised our Canadian friends that we will "do something" about acid rain. EPA has broad authority to limit air pollution, but the problem is really too big. The atmosphere over the United States now absorbs over twenty-five million tons per year of sulfur dioxide, and nearly as much of the nitrogen oxides, mostly in the Eastern states. There are regions of the East in which the average pH of rainfall is down to 4.0, and there is forest damage and damage to aquatic systems—lakes, streams, and fish. Unfortunately the cleanest coal is far away from this, in the West. No proposed technology for clean burning of dirty coal is particularly effective, though many are now in use and all are expensive. One can dream of removing 80 percent of the sulfur from coal emissions, but is more likely to achieve considerably less than that. And sulfur isn't the only culprit. Short of moving away from burning fossil fuels, there is no permanent solution. Even among the partial solutions none is known that doesn't somehow compromise our standard of living, or extract a high cost. It is when environmental concerns come into conflict with our lifestyles that our devotion is tested.

Lethal Pollution

The effluents from fossil fuel burning are also bad for human health, and sulfur dioxide has earned its share of credit for that problem too. The oxides of nitrogen, ozone (a form of oxygen), particulates, and other components present in smaller amounts, contribute to the attack on our collective lungs. The lung damage is likely to be the cumulative effect of exposure to smaller concentrations—one day in a smoggy or smoky city can irritate but is usually not lethal—but much of our information about lethality comes from a small number of extreme episodes.

We won't list them. The literature has many accounts of the sooty black palls that covered most of the large cities, London among the worst, as the Industrial Revolution spread the burning of coal. However, data about the effects on human health have been anecdotal until relatively recently; people just didn't collect the statistics. Even without detailed data, there is enough information to establish that major air pollution episodes kill some people. Most often the problem is an atmospheric inversion, of the type that causes routine smog in Los Angeles. Even for these dramatic events it is nearly impossible to unscramble the effects of the various products of combustion—everything happens at once.

Æsthetic arguments alone have persuaded many countries and cities to try to control atmospheric pollutants, so the worst of the pollution episodes of the middle of the century are unlikely to be repeated. Further, the main health problem is as usual the long-term exposure of large numbers of people to relatively small concentrations of lung irritants, so the familiar problem of data interpretation for small effects is present. And the health problems are not like cancer, where there is a long history of assuming linear dose-effect relationships. Instead, the pollutants cause or aggravate bronchial and other pulmonary diseases, and tend to speed up the deaths of the already infirm, or particularly sensitive. None of these episodes was so bad that healthy people dropped in their tracks, though some (like the one at Donora, Pa., in 1948) affected a large fraction of the population, and others (like London in 1952) were associated with so many deaths that there can be no doubt of their statistical significance.

It seems reasonable that some people are more sensitive to air pollution than others, and still others have personal habits (like smoking) that render them more vulnerable. Certainly it cannot be good to have emphysema and be in an irritating atmosphere. But if we want to be quantitative about any of these statements, we come up short of hard data. Many investigators have tried to do what the statisticians would call regression analyses—efforts to disentangle the effects of two or more causes—but the number of confounding variables is so great, and the data so sparse, that there is little agreement among them. For example, people of lower income tend to live in areas nearer the center of large cities, where the air is dirtier, but also smoke more. If their health is affected, is it air, poverty, or smoking? A regression analysis must account for these extraneous factors.

Given the complexity of the problem, the miracle is that the various estimates of mortality due to the burning of fossil fuels agree as well as they do. Not counting deaths due to the mining and transportation of coal or oil, there are numbers in the literature ranging from five thousand people per year in the United States to fifty thousand people per year, killed by air pollution from the burning of fossil fuels. Just the fact that the numbers are in that range, without being more precise, means that these effluents make fossil fuels by far the most deadly of our energy sources. And the radioactive contaminants in the smoke haven't even been mentioned.

The Greenhouse Effect

The summer of 1988 was unusually hot, not entirely from presidential campaign oratory. And it was not just 1988. The five hottest years since the middle of the last century, when worldwide temperature recording began, were in the 1980s. They were 1980, 1981, 1983, 1987, and 1988. That's impressive, and while it may be pure chance, it may not.

There are many effects that *could* have contributed to this, and many believe it was simply statistical fluctuation. It is not necessarily an early appearance of the greenhouse effect, which most experts would have expected in about thirty or forty years, but it could be. Either way, there isn't much that can be done about it now. So what is the greenhouse effect?

Gardeners are familiar with the uses of a greenhouse, a glass enclosure to protect fragile plants from the weather extremes, while admitting some of the light they need to grow. An unheated greenhouse is typically warmer than the outside, so the spring germination season comes a bit sooner than it does outside, and the fall growing season lasts a bit longer. In cold climates that doesn't prevent freezing inside the greenhouse, but it will always be warmer than the outside, especially on a clear day.

That all follows from a special and wonderful feature of light, its many colors. Light is electromagnetic radiation, which comes in wavelengths (the length of a single wave) ranging from the very short, epitomized by x-rays, to the very long, like the AM radio waves that bring us our mood music. There is electromagnetic radiation for every conceivable wavelength. Light, the electromagnetic radiation we can see, comes in a small range of wavelengths to which our eyes are sensitive. These are small wavelengths—two hundred thousand of them would fit across this page—but without them you wouldn't be able to read the page. The longer wavelengths of light are called red, about twice as long as the shorter ones, which are called violet. The rainbow spreads them all out for our enjoyment, violet on the inside, red on the outside, spread on a circle whose center always appears just below the horizon for an observer standing on the ground. And of course a pot of gold at the end.

Sunlight is mainly concentrated in the wavelengths we can see. That would seem a curious and remarkable coincidence, but of course it is not—we evolved in that light, and our eyes evolved to be able to see it. The chance of survival is greatly enhanced if you can see what is around you. The sun does emit some light with wavelengths too long for visibility—called infra-red—and some too short—called ultra-violet. The atmosphere blocks some of that light from the sun, fortunately in the case of the ultra-violet; that can produce sunburn or skin cancer while we try to get a perfect suntan. For better or for worse, the sun is our heater for our days on earth.

Unfortunately, it isn't quite up to the job. If the earth absorbed as much sunlight as it now does at this distance from the sun (93 million miles), and there were no greenhouse effect, the average temperature would be about twenty degrees below zero, Fahrenheit. That is not even survivable, let alone comfortable. The only reason the earth is habitable at all is that the atmosphere functions as a kind of greenhouse, exploiting the wavelength differences between the sun's radiation and that of the earth.

Yes, the earth also emits radiation, just as does every body in the universe, but its radiation is at much longer wavelengths, invisible to the human eye. The cooler an object, the longer the wavelength of its radiation. There are special infra-red films which will permit you to photograph the invisible radiation from a hot iron. Metalworkers and blacksmiths can tell how hot a piece of steel is by looking at its shade of red.

The infra-red radiation from the earth has wavelengths about twenty times as long as normal sunlight, and the atmosphere, just like the glass of the greenhouse, knows the difference. It lets most of the visible light in, while blocking the escape of the infra-red. This traps the heat we need, and keeps us warmer than the ordained twenty below zero, enough so to survive. The average temperature on earth is instead more like sixty degrees Fahrenheit, all because of the one-way transparency of the atmosphere. That's also how a greenhouse works, using glass instead of the atmosphere.

It's a delicate balance, with eighty degrees at stake. Of course there are hotter and colder places on earth, and unusually hot or cold seasons, and even unusually hot or cold years, but the average temperature of the planet stays reasonably constant over long periods of time. A couple of degrees can make a big difference to growing seasons for crops, moving the boundary for successful farming of a given crop toward the equator or toward the poles. In the midst of the last great Ice Age, nearly twenty thousand years ago, the average temperature was probably down less than ten degrees Fahrenheit, yet the glaciers were in our midst. The average temperature of the earth matters.

It's not the air itself that blocks the infra-red radiation, but a few minor ingredients, mainly water vapor and carbon dioxide. There are others, like ozone, methane, and the fluorocarbons, but they matter less. Ozone is famous at the other end of the visible spectrum, where it helps block the ultra-violet radiation from the sun, radiation that causes skin cancer. Ozone is being destroyed by other atmospheric pollutants, including the fluorocarbons, but that is another subject. We don't have much control over the water vapor in the atmosphere—four-fifths of the earth's surface is water, and it does what it pleases—but we do influence the carbon dioxide, and that's where the story really begins.

The normal atmosphere—the air we breathe—consists of about 78 percent nitrogen, 21 percent oxygen (the stuff of life), a bit less than 1 percent argon (an inert gas produced mainly through radioactive decay of the primordial potassium), and small traces of other gases. There is plenty of water vapor, and even visible water in the form of clouds, snow, and rain. The most plentiful of the less plentiful constituents is carbon dioxide, the other stuff of life. We animals live in a kind of companionable coziness with the plants on earth, a bit one-sided since they could easily get along without people. We breathe oxygen and give back carbon dioxide, while they take in the carbon dioxide and, with the help of sunlight, produce oxygen, and food for our tables. The carbon dioxide is continuously regenerated, as is the oxygen, and the arrangement works. It is not quite as simple as we pretended, because we are only one of many sources of carbon dioxide, while the plant life is the only source of oxygen.

The carbon dioxide does more than just feed the plants that feed us. It also helps block the escape of the infra-red radiation, to maintain that precious eighty degrees Fahrenheit. It matters how much carbon dioxide there is in the atmosphere, even though it is a minor component—it has a direct effect on the temperature of the earth. Not too much and not too little would be best. So how much is there, and how has it varied?

Not much is known about the distant past, but the amount of carbon dioxide in the air a hundred years ago was something like .027 percent, usually referred to as 270 parts per million, 270 ppm for short. It has been increasing ever since, and has just passed 350 ppm, with no end in sight. Why is it happening, is it bad, and what can be done? The answer to the last question is easy—very little. The answer to the next-to-last question is also easy—yes, it is bad. So let's begin with the first: why is it happening? At least that's a harder question.

A molecule of carbon dioxide, as the name implies, has one atom of carbon and two atoms of oxygen. The oxygen supply is plentiful—the atmosphere is 21 percent oxygen— so the key to carbon dioxide concentration is in the carbon supply, and that's easy to find. We Americans dump over five *billion* tons of carbon into the air each year, almost exclusively from burning fossil fuels. Every ton of coal burned to make electricity dumps nearly a ton of carbon into the air, every gallon of gasoline burned puts about five pounds of carbon into the air, every thousand cubic feet of natural gas burned discharges about thirty pounds of carbon, and so forth, almost all in the form of carbon dioxide. This is happening all over the world, and it is a prodigious amount of the stuff. More important, the dumping rate is increasing as the world becomes more industrialized and the need for energy increases accordingly.

In 1950 the United States was responsible for over 40 percent of the total carbon emissions for the whole world, but the world is catching up. Even though we now emit more per person than we did in 1950, we only account for about 20 percent of the total. In 1950 the Soviet Union contributed about a fourth as much as we, but has now nearly caught up. In 1950 China was tenth on the list, and is now third. Japan was ninth, and is now fourth. The newly industrializing countries need energy, most easily and cheaply obtained by burning fossil fuels as long as they last. Nearly every country is burning more fuel per person than it did in 1950, and the number of people has doubled. The world consumption of fossil fuels has been going up at an average rate of about 3.5 percent per year for the last thirty-five years. It adds up.

The increasing world population has contributed in another more indirect way, since people do have to eat and live; the forests of the earth are being cleared to make room for living space and agriculture. When asked to trade a present meal for a future concern for the climate, few hungry people will hesitate. This trend not only reduces the amount of vegetation needed to take the carbon out of circulation through photosynthesis, but any vegetation that is removed and not replenished turns itself into carbon dioxide. It doesn't matter to a tree whether it is burned or just decays, its carbon finds its way into the atmosphere in the form of CO_2.

Since we know approximately how much carbon is deposited in the atmosphere each year, and also know how much the carbon dioxide content is increasing, it is reasonable to ask if it's all accounted for. Does all the carbon end up just accumulating in the atmosphere, or are there natural processes that participate in a kind of atmospheric cleanup operation?

In 1958 a group anchored around David Keeling of the Scripps Institution of Oceanography set out to make regular measurements of atmospheric carbon dioxide at a site remote from local combustion centers. (It would have done no good to make measurements downwind of the Four Corners coal plant.) They chose the slopes of the Mauna Loa volcano, on the island of Hawaii, at an altitude just over 11,000 feet, to set up the Mauna Loa Observatory. Measurements have been made on a regular basis ever since, and form the most complete and continuous record of the steady increase of carbon dioxide in clean air.

The results are easy to express. The concentration in 1958 was about 315 ppm, and is now 350 ppm; measurements in other parts of the world, notably at the South Pole, are consistent with these numbers. Not only has the concentration been increasing, but its *rate* of increase has been increasing as fossil fuel usage has accelerated, almost as if the CO_2 were simply accumulating. The calculation is easy. The atmosphere weighs just about 5,200 trillion tons (about a million tons for each person on the planet), into which we are currently dumping 5.5 billion tons of carbon per year. Since a ton of carbon turns into a wee bit more than 3.5 tons of carbon dioxide (the oxygen weighs something too), we would expect the atmospheric concentration to go up about 3.7 ppm per year at the present rate, if the CO_2 were simply accumulating.

The CO_2 is actually increasing about half that fast, so only about half of the stuff we put into the air stays put, the rest somehow vanishing. Its destination is still something of a mystery, but most people point to the oceans as a reasonable sink. Certainly carbon dioxide can be dissolved in water (that's why we can enjoy both champagne and soft drinks), but the theory of just how the oceans clean the atmosphere is still incomplete. Even if half the carbon dioxide is being removed, half is staying in the atmosphere, and it is extremely unlikely that the human race will be weaned from the burning of fossil fuels very soon.

It's even worse than that. We've said that the world usage of these fuels has been increasing by about 3.5 percent per year, which would imply a doubling every twenty years. If that happened, and half the carbon continued to accumulate in the atmosphere, the present rate of CO_2 increase (about 1.5 ppm per year) would also double. We don't know the ultimate fate of the carbon well enough to be real prophets of doom, but we also don't know that this won't happen. We are doing a giant experiment with the earth, and the stakes are high.

If the CO_2 does accumulate as it has been accumulating, and the world population continues to grow and increase its hunger for fuel and for arable land, the inevitable conclusion is that the carbon dioxide concentration in the air will double some time in the middle of the next century. It could come sooner, and it could come later, but come it will. So what are the likely consequences?

That would seem like an easy question; if the carbon dioxide blocks the escape of infra-red radiation, and the amount of carbon dioxide increases, it should do a better job of keeping the heat in. That would mean the earth would get warmer, growing seasons in the middle latitudes would get longer, the sometimes miserable inhabitants of the northern part of our country wouldn't need to go to Florida or California in the winter, and everyone would rejoice. Of course it isn't that simple.

Hot air tends to rise; it is lighter than cold air. Hot air balloons will stay aloft as long as they are kept hot, but will fall when they cool off. The hotter air near the earth's equator also rises, while the colder air near the poles tends to sink, and a circulation or convection pattern develops in which warm air rises near the equator, journeys north, then sinks, and finally is carried south to complete the pattern. The result of all this is that solar heat is transported from near the equator, where it is most plentiful, to the polar regions that are most in need. It is a kind of thermal charity organization.

But it isn't even that simple, because the earth isn't sitting quietly while all this is happening. It is also rotating on its axis, approximately one full turn per day, and this rotation confuses the wind pattern. The effect is complicated—it is called the Coriolis effect—but it is familiar from observation of ice skaters or divers. They spin faster when they curl themselves up into tighter balls. Since the earth is more or less round, winds in the northern hemisphere tend to be deflected to the right as they move along. If it sounds too complicated, trust the author.

The result of all this interplay of heating, convection, and rotation is that there is not a simple wind pattern of high winds heading north and low winds heading south, but a complex pattern of trade winds, doldrums, prevailing winds, etc. This is called the global circulation, providing gainful employment for untold numbers of meteorologists, mathematicians, and computers. It is still not well understood, because there are many more factors than have been mentioned. The computer models that deal with this are called GCMs, for global circulation models. There are several good ones, which frequently agree with each other.

There are other effects. If the earth were to warm up, more water would evaporate from the oceans, more clouds would form, and more direct sunlight would be reflected before it ever got to the surface. This is a cooling effect, which might somewhat counterbalance the heating effect. It is unusually hard to calculate theoretically. One has to start with the radiation coming to us from the sun, and trace all the significant paths through which it is reflected, absorbed, reradiated, and finally disposed of, whatever its ultimate fate. To do a good job of dealing with the consequences of a change in one component of the atmosphere, all these effects must be estimated, and it is a miracle of modern computing that it can be done at all.

The GCMs in use nowadays do a pretty good job of calculating the effect of a potential doubling of the carbon dioxide content of the atmosphere, but more research is truly needed. (Music, as was said before.) The details of the impending changes of climate are still beyond our grasp, though the broad outline is clear.

Four major groups work on such calculations in the United States, and one can get some idea of the accuracy of the models by asking how well they can account for the present climate. The answer is that they do quite well for the global average temperature, within about one degree Celsius (nearly two degrees Fahrenheit), but not nearly as well in describing more local weather, often missing by five degrees Celsius or more. The larger the scale the better the results.

All models agree that the net effect will be a general and global warming of the earth; they only disagree about how much. None suggest that it will be a minor effect, to be ignored while we go about our business. The estimates vary everywhere between three and five degrees

Celsius (five to nine degrees Fahrenheit), and those are enormous changes. Recall that it was about five degrees Celsius cooler during the last Ice Age; far less is known about what would happen for a comparable warming.

There will certainly be a major effect on agriculture and ecosystems, and some areas will doubtless benefit from a moderation of their climate. Some barely habitable areas will become even less attractive—the number of 100° F days in Washington, D.C., may well zoom, forcing us to move the capital north. Either that or only masochists and nincompoops will serve the government. Rainfall patterns may change enough to make good farming areas bad, and bad ones good. Or perhaps the climatic effects will be tolerable. We simply cannot now predict that well—it is a global experiment.

One additional effect has been widely publicized. Everyone knows that ice melts in warm weather, and both the northernmost and southernmost parts of the earth are covered with ice. What will happen if much of it melts? Fortunately, much of the northern ice is simply floating on the water, and its melting over a long time will hardly affect sea level. Ice that melts while floating in water doesn't make the water rise. To see this, fill a glass of water to the brim, with an ice cube floating in it and therefore poking above the brim. When the ice cube melts, the water will still be at the brim, with no overflow. That's why icebergs and ice cubes protrude above the water—ice is less dense than water.

Some of the ice, however, is not afloat, but is packed on solid earth, and that icewater will flow into the ocean. The major northern ice cap is on Greenland, not as large as it looks on a Mercator's projection, but still large. In the south, Antarctica is a solid continent covered with deep ice that has accumulated for centuries. If all that ice were to melt and find its way into the oceans, it has been estimated that sea level would rise by some tens of feet, but that is an extremely worst-case picture. More realistically, the level of the ocean has in fact been rising all through this century, at about one millimeter per year, adding up to a few inches in the century. This rate could triple or more, so that sea level by the end of the next century could easily be several feet higher than it is now. The social impact of such a rise in sea level, if it happens, will be visited on generations later than ours, but not much later. Coastal cities would be inundated, shorelines would recede, and so on. Not the end of the world, but not a pretty sight.

Other greenhouse phenomena have been omitted: the effects of the other atmospheric gases that absorb infra-red light; the increase in sea level just from the fact that the warming oceans will expand, as most things do when they get warmer; and so forth. Yet, despite the complexity, the bottom line is that the earth will be substantially warmed by the accumulation of man-made gases, mainly carbon dioxide, and that warming could conceivably approximate the climate at the time of the dinosaurs. It seems likely, but not certain, that sea level will rise accordingly, conceivably by several feet or more. We are doing this to ourselves.

Can anything be done to slow it down? The only option in the long run is to decrease the amount of waste gases deposited in the atmosphere. That would require global cooperation and sacrifice now, to avert something far in the future, and a conjectural something at that. There is no evidence in human history that that is in the cards, but one can always hope. There has actually been some progress in one atmospheric area in which not much

sacrifice is involved: the chlorofluorocarbons used in re-
frigerants and spray cans are deadly to atmospheric ozone,
and they are being phased out in favor of more degradable
chemicals. But there is no evidence whatever of any worldwide willingness to reduce the rate of deforestation or to
reduce the consumption of fossil fuels. There are simply
no substitutes for most uses.

Of course, for electricity there is nuclear power, which
produces no greenhouse gases at all, but which has political
problems. We have mentioned that France is now producing about 70 percent of its electricity with nuclear power,
higher than any other country in the world, and therefore
qualifies as a good world citizen. The United States, the
world's worst citizen in dumping carbon into the atmosphere, has also stopped its nuclear program in its tracks.
It has been amusing to watch the reaction of the professional anti-nuclear folks as the specter of the greenhouse
effect has loomed more visibly. It obviously produces an
argument in favor of nuclear power, which is intolerable
to many of them. The reactions have been like those of
people who are told they have an incurable disease—first
denial, then anger and resentment. In the normal sequence
this should be followed by acceptance, but that hasn't yet
happened.

A greater shift to nuclear power would somewhat delay
the effects of global warming, as would a switch to a less
profligate way of life (some people call this conservation),
and there is still room for improvement there. It would
also help a bit to use more natural gas, since that releases
less carbon. But there is nowhere in evidence any effective
solution to the problem, on the scale necessary to really
matter. Even if the world stopped adding carbon dioxide to the atmosphere tomorrow—a pipe dream—there is

already enough to affect the climate. Most likely the recent string of hot years is just a climatic fluctuation, not the emergence of the greenhouse effect, but one cannot be sure.

If this were a science fiction novel, our hero would now discover a way to make energy without waste, at the same time learning to feed and house the world while replanting the forests. We might as well throw in a solution to the population problem. But this isn't a science fiction novel.

17
Nuclear Winter

FACTS AND UNCERTAINTIES

It seems only natural that this chapter should follow a section on the greenhouse effect. That had to do with the climatic effects of the extra greenhouse gases we are dumping into the atmosphere. They reduce the earth's ability to send back into space the heat we receive from the sun, and will ultimately make the earth warmer. This chapter is about blocking the original solar radiation, thereby cooling the earth. Both are attributable to technology. Though our current concern is that nuclear war might lead to solar obscuration, nature has done it in the past, without our help.

On the rim of the Pacific Ocean, where some of the great tectonic plates that form the crust of the earth grind against each other, volcanoes are common. Islands have been created and destroyed by volcanic eruptions—the Hawaiian Islands were formed that way—and the events can be violent indeed. There are volcanoes more or less everywhere near the plate edges, but the largest concentration is in Indonesia, where nearly a hundred volcanoes have been known to erupt in historic times. The Pacific Rim has been called the Ring of Fire, and has three-quarters of the world's 800-odd active volcanoes.

A volcano can erupt relatively quietly, affecting only the local area, which may experience either a rain of volcanic ash or a river of molten lava, or both. At the other extreme the eruption may fill the sky with debris that migrates to the far ends of the earth. The largest amount of lava that is known to have been produced in a single eruption is a few cubic miles (the volume of Crater Lake, in southern Oregon, which was formed in an eruption about 7,000 years ago). The amount of debris put into the air can be comparable, and can partially block the sunlight all over the world.

The best-known volcanic eruption of modern times (thereby excluding Vesuvius, which buried Pompeii under 50 feet of ash) was in 1883 on the island of Krakatoa, on the Sunda Strait, in Indonesia. It almost completely destroyed the island, removing the top and leaving the base a thousand feet below sea level. The ensuing tidal wave (a misnomer for a tsunami, which is somewhat less of a misnomer) killed tens of thousands of people on the neighboring islands. The atmospheric debris spread around the world, producing beautiful sunsets for about two years, but with no noticeable effect on the weather.

Sixty-eight years earlier, in 1815, the Tambora volcano on the island of Sumbawa had had an even more violent eruption, quite possibly the largest in several centuries, again with the debris encircling the world. The following summer was unusually cold throughout Europe (there is a Byron poem to mark the occasion), and some old Chinese records that have recently been found indicate that it was also unusually cold in China. That year, 1816, has come to be known as the year without a summer. Whether any of these global weather manifestations, other than sunsets, resulted from the volcanic eruptions is another matter. The current weight of expert opinion is that they did not, but the quantity of debris that a volcanic eruption can put into the atmosphere is formidable and, as we will see, comparable to the amount that would result from a major nuclear war. But it is different stuff.

The other known natural event that produced effects of this sort was the impact of a large extraterrestrial object some seventy million years ago, which seems to have spread so much debris in the atmosphere that it caused widespread extinction of animal species, including the dinosaurs. There are still a few doubters that this actually happened, but the evidence is quite conclusive. So natural phenomena can affect the opacity of the atmosphere, and there is good reason to believe that a large enough event can affect the climate. There is no magic here, just the complex heat balance of the earth, as discussed in the last chapter in connection with the greenhouse effect.

In the event of a major nuclear war—we and the Soviets each have tens of thousands of nuclear weapons—there would surely be enormous global fires spewing smoke and dust into the atmosphere, substantially affecting the delicate radiation balance that preserves the earth's temperature. The idea was first published by Paul Crutzen and John Birks in 1982, promptly christened "nuclear winter," and it is now estimated that something like a hundred million tons of smoke would be produced in the first few days after a nuclear exchange (a polite term for a nuclear war). Much of this would be in the form of small particles of carbon (soot) which could be effective blockers of solar radiation, much better at it than volcanic ash. Black smoke can be pretty black.

Before getting into the technical details and uncertainties, some side comments about the importance of this subject are clamoring for expression. The next two paragraphs contain some personal views of the author (as, of course, does the rest of the book), but those views are not uncommon among experts.

We have known since the beginning of the nuclear age that nuclear weapons could cause great fires—it happened in 1945, on the only occasions, so far, in which nuclear weapons have been used in anger. The new idea in 1982 was that the fires produced in a war among the major powers could be so large that they could have global impact, far beyond the already frightful effects of nuclear war. Some extreme views were that such a war would wipe out all life on earth, or at least all human life, but those views were based more on a quest for political effect than on science.

The subject then attracted some well-known names in the scientific community, who popularized it as still another reason to avoid nuclear war, and to work toward nuclear disarmament. This author (who has chaired one government committee on the subject) was always puzzled by the intensity with which these views were expressed, almost as if the speakers had just discovered that nuclear war would be bad. He doesn't know a single person, anywhere, who is in favor of nuclear war, yet the oratory seemed to be directed at such people. Many have spent their lives at the serious and responsible job of preventing nuclear war, and can easily find it offensive to have the opposite suggested by well-intentioned public figures who have just discovered a cause. The prevention of nuclear war, second only to overpopulation as a real and immediate threat to the human race, requires work, not bombast. The record of the last forty years shows that we have so far been successful at preventing major wars. As with the other risks discussed in this book, that is no cause for complacency. End of lecture.

Fortunately, the problem of understanding the phenomenology of nuclear winter is similar to that of understanding the weather, the climate, and the greenhouse effect, so an experienced community of first-rate scientists is available to study the subject. The rest of the discussion will be based on their work.

The scenario is simple enough, though it is obviously a "worst case." It is that for some reason the inhibitions that have prevented nuclear war among the great powers for all these years break down, and a massive nuclear exchange takes place between the United States and the Soviet Union, the only nations well enough armed to produce global effects. It doesn't matter to the scenario whether others join in; this is bad enough. Calculations (and what little experience there is) show that the major cities under attack would burn, and perhaps the forests and much of the brush. There would be immense fires, on a scale unprecedented in human history. Forest fires are common enough, but simultaneous fires over a substantial fraction of the earth's dry land are not, and civilization has accumulated large stocks of combustible materials in the cities and elsewhere. At this moment the United States has in stock around the country some four hundred million barrels of petroleum and its products (a barrel is defined in the oil business to be forty-two gallons), corresponding to about seventy gallons for each of us. Most of this would probably burn, as would the comparable supply on the other side. There are four automobile and truck tires per person in the United States, many of which would burn. And so forth. That already adds up to over fifty million tons of combustibles, and combustible forests and farmlands easily bring us up to the estimate of a hundred million tons of smoke.

If that amount of smoke were to released in a short time, and a comparable amount on the other side of the ocean, it would surely darken the sky and turn day into night. It is a huge amount of material, though we should maintain some perspective by recalling (from the last chapter) that over fifty times as much carbon is deposited in the earth's atmosphere from the burning of fossil fuels *each year*. However, that is in the form of the visually transparent gas carbon dioxide, not black smoke.

The effective response to this threat is obvious—avoid nuclear war—but of course that was true without nuclear winter to provide extra motivation.

The early calculations of the effects of this much smoke injection were what are called one-dimensional, which is to say that they didn't treat the spreading of the smoke very well. Nor did they treat other mitigating effects, like washout by rain. Partly because of these oversimplifications, the calculated effects were dramatic, predicting that the darkness would last for weeks or months, circle the globe, and produce temperatures well below freezing in summer in the middle latitudes. This is where most of the world's population lives. Such temperatures would wreak havoc on the ecosystems that feed us, though the local supermarket might have still other reasons to disregard regular hours right after a major nuclear war. The ultimate disaster of a nuclear war would be even more ultimate.

The nuclear winter phenomenon is real, the remaining unknowns requiring better calculations. Despite what may have been inferred from the comments made above, we should try to learn as much as possible about the subject, if only to contribute to better decision making. It was therefore obvious after the 1982 debut of the subject that more research was needed, both through better calculations and through research on the phenomenology of large fires. We can use the large forest fires which are endemic in parts of the United States, and also the large controlled burns deliberately set to burn off the fuel that might later feed uncontrolled fires. (There has been some outcry about the latter, some people having been fooled into believing that the fires were being set solely to study nuclear winter.)

One of the less understood features of large fires is the generation of so-called firestorms, events in which the fire is so big that the rising hot air from the fire sucks in the surrounding ground-level air, both feeding the fire and fanning the flames. The condition can sustain itself until there is nothing left to burn, and control is generally impossible while it is going on. While an ordinary big fire may possibly burn everything in sight, a firestorm almost certainly will.

A fire built in a fireplace has, on a small scale, many of the characteristics of a firestorm. The hot air that rises through the chimney creates a draft, drawing fresh air in at the base of the fire, which then provides the new oxygen the fire needs. Everyone knows that a good draft is important to a good fire, and of course the air flow also keeps the smoke out of the house. A firestorm makes its own invisible chimney in the atmosphere, and the winds that flow into the bottom can exceed hurricane force. The smoke from a firestorm can be deposited high in the atmosphere, where it can both spread furthest and last longest.

It's not easy to predict when a fire will cross the invisible threshold and turn into a firestorm. Many large cities have burned in the past—cities are full of combustible materials—London in 1666, Chicago in 1871, San Francisco in 1906, Tokyo on several occasions, and dozens of others. It is hard to tell from the memories of witnesses which of these, if any, was a full-fledged firestorm. Probably some were.

During the World War II bombing of cities it was noticed that more destruction—the purpose of bombing is destruction—was caused by the fires than by the general-purpose bombs. The air raids were therefore soon aimed at setting fires, with a firestorm a possible bonus. The German raid on Coventry was about 25 percent incendiary bombs, the raids on Cologne and Dresden were 40 percent and 75 percent respectively, and the major American raid on Tokyo was 100 percent incendiary. Firestorms were sometimes started, wreaking enormous damage, and sometimes not. The raid on Hamburg (25 percent incendiary) caused a firestorm that generated hurricane-strength winds. An air raid is bad enough without a firestorm, but a firestorm makes the damage much more complete, the inward winds confining and feeding the fire. A firestorm can also deposit the smoke as high as the stratosphere.

Unlike conventional bombs, nuclear explosions can always start fires. The explosion itself is hot, far hotter than the sun, and exposure of combustible materials to the radiation from the bomb can start fires at great distances. For a one-megaton bomb, a fire can be ignited at a distance of about five miles. The only direct nuclear experience is again that from World War II. In Nagasaki there was a fire but no firestorm, while at Hiroshima there was a firestorm which began a couple of hours after the explosion, lasted about twelve hours, and burned out what was left of the center of the city, between four and five square miles. A major nuclear war would certainly produce fires on a prodigious scale, and at least some of these might well turn into massive firestorms. On top of that, there is nuclear winter.

Given the standard scenario, the predictions for the amount of smoke injected into the atmosphere are generally right. That doesn't mean right to within 10 percent, but does mean that it would be unreasonable to expect that the estimates are ten times too high. This is for the arbitrary scenario of complete exchange of nuclear weapon inventories, which will be pursued to the end.

Once the smoke is deposited in the upper atmosphere it will surely darken the local area, and will just as surely spread around the earth, as did the volcanic ash from Krakatoa. The time it takes to do this depends on where in the atmosphere the smoke is deposited, but is something measured in weeks. The great unanswered questions are whether the cloud will persist that long, be mixed with enough clean air to make it an ineffective solar blocker, be scavenged by rain, form clumps and fall to the ground or ocean, or suffer some other fate.

The answer isn't very well known—this is a truly complicated problem, taxing the best computers and wisdom—but the calculations are being continually refined to take into account more and more of what is known about the atmospheric circulation and about smaller-scale atmospheric phenomena. As the knowledge accumulates, the expert predictions are leading to smaller and smaller predicted climatic effects. They are still alarming, because the underlying theory of the effect is surely right, but the phenomena are not as simple and the effects not as large as appeared from the early one-dimensional models that generated all the public attention. Some experts have suggested that the nomenclature be changed from nuclear winter to nuclear fall, because the temperatures now predicted for summer in middle latitudes are more like autumn than winter.

And that's the story of nuclear winter at this time. It is real, but one of the less important reasons to work at the prevention of nuclear war.

18

Non-Ionizing Radiation

FACTS AND UNCERTAINTIES

It was said in Chapter 16 that electromagnetic radiation comes in all colors, the wavelengths running all the way to x-rays and beyond on the short side and ordinary radio waves and beyond on the long side. There is no limit in principle in either direction. There are radiations from nuclei and in the cosmic rays whose wavelength is much shorter than ordinary x-rays, and there are radio waves far longer than those of the commercial broadcast bands. The wavelength of any of these is inversely related to what is called its frequency. A commercial radio station whose frequency is a million cycles per second (called hertz and abbreviated Hz), 1000 or sometimes 100 on your AM dial, has a wavelength of approximately a thousand feet. A hundred million cycles per second (100 MHz on your FM dial) would have a wavelength of about ten feet, and so forth. Radiation emitted by the 60 Hz electricity that powers most of our homes (50 Hz in some countries) has a wavelength over 3,000 miles, and it would take only a few wavelengths of any radiation emitted by our brain waves or heartbeats to reach from here to the moon.

There is no magic distinction between short waves and long waves—they all belong to the same family. Still, it is useful to distinguish between those wavelengths that can cause ionization, and thereby damage human tissue, and those that cannot. As a general rule, not to be taken too literally, radiation on the short side of the wavelength of visible light is ionizing. Even ultra-violet light can cause skin cancer, and most people, especially those with fair skins, are now warned by dermatologists to stay out of the sun during the summer. (This is one reason to be concerned about the ozone layer in the atmosphere—ozone absorbs just that part of the ultra-violet light that causes both sunburn and skin cancer.) As we move into the even shorter x-ray wavelengths the cancer potential is well established, as discussed in Chapter 15. Radiation with a longer wavelength than the infra-red end of the visible spectrum is non-ionizing, but can be damaging in other ways.

Before the beginning of this century we didn't know much about the effects of non-ionizing radiation on the human body; radio was only invented in the last few years of the nineteenth century. It was still not possible to generate radio waves of sufficient power to affect humans. Even the light bulb or incandescent lamp, which is one way to convert electrical energy into visible radiation (light), was only invented a bit more than a hundred years ago. The first use of electricity itself on the human body was in 1908, when it was realized that our nerves were too slow to keep up with a sufficiently high frequency electrical current; that made it possible to pass high currents through the body without shock. So diathermy machines were invented, to warm the body without shocking it, and thereby promote healing. We still use heating pads for this purpose, or infra-red heat lamps, without passing electricity through the body. Diathermy machines are also still in use.

It was during World War II that radar came into general use, and it was then noticed that anyone who stood near the beam of a high-power radar set could begin to feel uncomfortably warm (or comfortably, depending on the climate). It has been claimed that microwave ovens were invented when an engineer passed through the beam of an operating radar set with a chocolate bar in his pocket. According to the legend, the bar melted and microwave cooking was born. If the story is true, and it may be, it will not mark the first time a stroke of genius was traceable to a chocolate bar.

In any case, sufficiently powerful microwaves can warm the body, just as they can warm a rib roast, and one would have to be crazy to put his hand into a microwave oven when it is on. For reasons of safety all such ovens are required to have an interlock—an automatic switch that turns off the power when the door is open. It wouldn't take a hand very long to suffer real damage.

Just to limit the unintended heating effects, there is a federal regulation which limits the emissions from microwave devices to less than $5\,\text{mw/cm}^2$ (milliwatts per square centimeter, a measure of intensity of the radiation), at which point there could be some warming of body tissues. In most cases moderate warming would do no harm—fever, for example, is one way our bodies fight disease—but there are some particularly sensitive tissues that don't have good self-cooling mechanisms (the eye is one). The Soviets have far more stringent standards for exposure to non-ionizing radiation, because they believe that there are psychological effects beyond mere warming, and that these are more important. It has never been possible for us to verify the Soviet claims of the existence of these effects, so there is no compelling reason to protect against them, but there is a persistent group of reputable scientists in the United States who believe that there are subtle effects of which we are unaware. As in many such cases, the reputable scientists have been joined by enough nuts to compromise the quality of their efforts. If there are such effects, they are small and elusive. In all science the test of a claim is whether another scientist, without a stake in the outcome, can duplicate the effect. Where this has been tried, the results have usually turned out badly for the claim that extremely low levels of radiation are damaging to people.

It is important to keep clearly in mind that an effect per se is not necessarily a harmful effect. It is easy to demonstrate that a powerful radio beam can produce a sensation of warmth, but that isn't necessarily harmful. It is easy to demonstrate (we do it in elementary physics teaching and nature does it in thunderstorms) that a large enough electric field can make your hair stand on end, but that isn't necessarily bad. There is a marvelous set of true stories that some people who lived near high-power AM radio stations were hearing music in their heads, which was finally traced to the rectifying effect of the fillings in their teeth. Depending on the programming, this could be damaging to mental health. And so forth. The point is that even when valid research leads to claims of an "effect," it doesn't follow that the effect is bad.

There are two important dependencies on frequency or

wavelength. We have already mentioned the relationship between frequency and wavelength—they are inversely related.

In the low-frequency domain (not x-rays, which are different in this regard), the degree to which the radiation can penetrate the human body (or a rib roast, for that matter) depends on its frequency or wavelength, and is greater for longer wavelengths. Thus infra-red radiation hardly penetrates the body at all, though an infra-red lamp will produce a feeling of warmth by warming the skin, while the radiation from 60 Hz house wiring penetrates completely. For microwave ovens (the term "microwave" means having short wavelength), there is a federal rule standardizing the frequency at 2450 MHz (a wavelength of about five inches), and at that frequency the penetration depth is about an inch. That is why the inside of a large block of frozen food thaws more slowly than the outside. In fact, the main difference between ordinary baking and microwave cooking is that oven baking is done entirely from the outside of the food, while the microwaves penetrate about an inch or so, and deposit the heat there. This so-called penetration depth is an important feature of non-ionizing radiation.

The second dependency arises from the fact that all radiation, whatever the frequency, is ultimately generated by oscillating electric currents. Sometimes the currents are in wires, sometimes in the interior of atoms, and sometimes in the nuclei themselves, but they are always there. Sometimes a wire is deliberately used to produce radiation by sending an electrical current through it—then it is called an antenna. Other times the radiation is unintentional, but oscillating currents always emit radiation. The frequency of the radiation is the same as the frequency of the current; if the current goes back and forth a million times a second, the radiation will have a frequency of one megahertz, 1 MHz.

Radiation is much more efficiently produced at high frequencies than at low, so the radiation from 60 Hz house wiring is weak indeed. On the other hand the gadgets (normally magnetrons) that produce the radiation for our microwave ovens are quite efficient.

Some of the early concerns about low-frequency non-ionizing radiation appeared in the early 1960s, in connection with the problem of communication with nuclear submarines in the deep ocean. The strategy of deterrence, on which both we and the Soviet Union have long based our strategies for the avoidance of nuclear war, requires that neither side be vulnerable to a surprise attack by the other. That doesn't mean the attack couldn't take place—each of us has more than enough nuclear weapons to destroy the other—but only that there is no way to get away with it without suffering the consequences. This requires that each side have a relatively invulnerable supply of retaliatory weapons, in sufficient numbers to make it clear that retaliation in kind could and would take place. Some people are horrified by this so-called balance of terror, but it does seem to have worked for over forty years. The underlying arguments are not relevant here—many learned books have been written—it is only necessary to note the perceived need for the invulnerable deterrent force.

For the United States, the force has traditionally been formed out of what is called the triad, a combination of aircraft, land-based missiles, and submarines, so that no aggressor in his right mind would imagine that he could destroy them all at once. Critics have wondered whether we need all three, but that is again not the subject of this book. In any case, for a deterrent to be credible—and credibility is all that is required for deterrence—it must also be possible to communicate an order to fire when the occasion demands. If the other side knew that the retaliatory force would never receive the order to retaliate, the deterrent wouldn't deter. The whole complex structure of command, control, and communications is called C^3, C-cubed. It is in principle easy to communicate with the air forces, which spend most of their time on the ground anyway, and a very robust system has been developed to assure communication with the missile silos; the submarines pose a harder problem.

A submarine achieves its security not by being in a sturdy concrete silo, like a missile, nor by taking to the air if there is mischief afoot, like an airplane, but simply by hiding in the vast oceans. The oceans are so large that hide-and-seek is usually won by the hider—a Navy acre is a whimsical term for a million square miles of ocean. Gargantuan sums have been spent by both sides on efforts to discover how to find hidden submarines in the open ocean, and if there has been any success it too is well hidden. The command and control problem follows immediately: if a submarine is that well hidden, how can anyone communicate with it? Without communication it is not a very good deterrent.

By radio, you might say, and that brings us back to the subject of the chapter. Unfortunately, the oceans are made of salt water, very similar to the fluids in our bodies (remember that despite what the creationists say, humans evolved from seagoing creatures, and our bodies remember in their own way). Therefore, if a commander tried to communicate with a submerged submarine at, say, a frequency of 2450 MHz, the message would penetrate the water to a depth of about an inch, just as that frequency does in a roast in a microwave oven. That won't do much good for a submarine that may be hundreds of feet below the surface.

But the penetration of radio waves into the ocean varies as the inverse square root of the frequency, which means that if the frequency of the radiation is divided by a factor of four, the radiation will penetrate twice as far. A simple calculation shows (taking into account that our bodies aren't really exactly like seawater) that a frequency of 60 Hz, like the power-line frequency, has a penetration depth of about a hundred feet. If that frequency were used for communication, some radiation would get down to a reasonable submarine operating depth, and at least one-way communication would be possible. So a project was born, more than twenty-five years ago, to communicate with submarines at extremely low frequencies, and therefore extremely long wavelengths, just to penetrate the seawater. It has had many names over the years, starting as Sanguine, taking the alias of Seafarer and moving from Wisconsin to Michigan when the going got rough, then masquerading as plain old ELF (for Extremely Low Frequency), and now just hibernating with a token deployment, well short of operational utility.

Recall that low-frequency radiation is hard to generate efficiently, more so as the frequency decreases, and this is an extraordinarily low frequency for radio communication. The plan was originally to string thousands of miles of antenna wire through northern Wisconsin (there are technical reasons why it is good to be over pre-Cambrian rock), and run thousands of amperes of electrical current through the wire. It was at that point that environmental and health concerns erupted, since these wires would be going through good dairy farmland, and it was easy to believe that they would do harm. Certainly the Wisconsin delegation in Congress was against it—the economic health of Wisconsin is closely linked to dairy farming.

Though it would be easy to attribute the doom of this grandiose project to the health concerns, it was at least as much due to the fact that the Navy, for whom it was intended, really didn't like the idea in the first place. Just as the Air Force resists any idea that forces it to spend money on projects that don't provide more airplanes for pilots to fly, and NASA de-emphasizes projects that don't provide vehicles for astronauts, the Navy doesn't like to use resources on projects that don't provide ships to sail and command. The Army, not to be different, used to be devoted to sabers and horses, which is why tank regiments are still called by their old cavalry names, but was unable to continue to justify them for combat use. (National security and welfare are insignificant compared to these primal instincts.) Without the dedicated backing of its Service, no project can survive.

The ELF issue was the precursor to the current battle about power lines. In the end, the Navy supported large research projects to see whether high-power radiation at a few tens of hertz did in fact have any effect on health, and none were ever substantiated. Cows were exposed to the fields with no effect on milk production, and people, insects, other animals, and plants alike showed no alarming effects.

Fanciful stories abounded at the time. One that was especially popular was based on the fact that some of our brain waves (the so-called beta waves) are at frequencies close to those that were proposed for the communication system, and the alpha waves have just about half that frequency. Though the magnetic fields (most important at short distances) were weak, it was thought that they could somehow synchronize our brain waves and alter our minds. The Soviet reports from the early 1960s about psychological effects on people who worked near high-voltage fields seemed relevant. Stories circulated in which flickering lights at just this frequency drove people mad, along with legends of astronauts being affected by rocket vibrations at such frequencies. None of these was ever corroborated, but they were interesting enough to be entertaining and believable, and in the end contributed to the ultimate demise of the project. What didn't die was the concern about low-frequency fields.

There has long been a small biomedical community which believes that very small electromagnetic fields have potentially harmful effects on some fundamental life processes, including brain function. The people who hold these views do have some fragmentary evidence of effects at fields below the level at which one would expect anything at all to happen. The evidence is far from clear, but it suggests that we don't know everything about the subject—hardly a novel position. This group of dedicated people has never been able to demonstrate any actual, as distinguished from potential, *harmful* effect of these low levels of radiation on people, but they are still trying. Public policy in the area is therefore based to some extent on conjecture. Recall that the Soviet limits on microwave exposure to the population are far more restrictive than ours, but without any persuasive evidence we don't have. They also make exceptions for military programs, which don't have to obey the rules. The microwave standard in the United States is based almost entirely on body heating by the radiation, again because no deleterious effects have ever been demonstrated at lower levels. But power-line fields are not microwaves.

There has been a new flurry of interest in recent years, because the steady growth in the use of electricity (a few percent per year, which adds up over the years) has led to the need for new power lines through an increasingly crowded country, and improvements in technology have led long-distance power transmission to higher and higher voltages. Partly because no one wants a power line in his backyard, and partly because of increased environmental awareness and activism in the country, few power lines are built without local opposition. As people seek weapons to use in opposition to the lines, it is natural to turn to the potential health problems they might bring. There is nothing intrinsically wrong in doing so, since it is an area that still harbors some uncertainty. So what are the technical issues, and the evidence?

By a large margin, the most solid recent study of the effects of power-line fields was supported (to the tune of $5 million) by the New York State Power Authority and a consortium of New York utilities. It followed persistent reports that people who live near power lines (the lower-voltage distribution lines in and near cities) either have a higher than normal rate of childhood leukemia, or more miscarriages or longer gestation periods (electric blankets were fingered as the culprit for these), or suffer from other problems. Every one of these reports was based on a human epidemiological study which suffered from some kind of statistical or structural defect, but couldn't be discounted out of hand.

The study was a responsible one, with a competent scientific advisory panel, and the conclusion, in a nutshell, was that all of the evidence had alternative explanations, or was uncontrolled, or couldn't be duplicated, or suffered from bias, or was in some other way unconvincing. In one case the data were collected by phoning people who had published birth announcements, thereby automatically selecting an upper socioeconomic group. In another study which reported changes in reproductive patterns, the children had been born long before the subjects worked in an electromagnetic environment. In still another, the classifications were made subjectively by a person who knew what result was wanted. (This doesn't suggest that he was dishonest, but in a good experiment one guards against unintended bias.) When an effort is made to distinguish between homes near power lines and homes far from them, it is obvious that there is a social bias—rich folk don't live near power lines. And so forth.

The bottom line of the New York study was that there is reasonably clear evidence that low-frequency electromagnetic fields can affect cells in a variety of ways, but little evidence that there is damage at relevant levels, or that the current standards are inadequate. There is no unflawed epidemiological evidence of deleterious effects of power lines on the population, though there are enough flawed studies to keep the subject interesting. In short, we know a great deal, but don't know everything, so more research is needed.

It is appropriate here to mention a splendid colloquium that was delivered a few decades ago by Irving Langmuir. He spoke about scientific error—the discovery of effects or things that aren't there—and sought to find a common pattern by examining some famous historical cases in which great discoveries were announced, were believed for a while, and then turned out to be wrong.

The cases he studied did indeed have some common

features. First, the scientists involved were never deliberately dishonest—each devoutly believed that he had made an epochal discovery. None of it was fraud in the sense that we speak of bank fraud. (Langmuir entitled his colloquium "Pathological Science," presumably in recognition of the fact that he was dealing with a disease, not a felony.) The purported discovery was always just barely detectable above the background of confounding phenomena, so that when others tried and failed to reproduce the experiment, it could always be said that they hadn't done it exactly right. The discoverers always resented the "orthodox scientific establishment," which didn't give their work the credit it deserved. There is more, but the point is clear. Even an honest scientist is capable of self-deception, and that can lead to deception of others. That is why science thrives best in an atmosphere of open communication, and in no other environment. It is also why good scientists try not to resent criticism—they know that they are sometimes wrong, and that self-criticism is somehow never as effective in separating truth from fantasy as mutual criticism among peers. That they be peers is essential—the accuracy of the theory of relativity cannot be determined by a vote of the legislature or of a school board. To say nothing of evolution.

Part 3

Coda

19

Just Enough Statistics and Probability

If this chapter were a bottle of wine or a package of sauer-kraut, the consumer protection laws would require a label: WARNING, CONTAINS MATHEMATICS! Mathematics, like the carcinogenic chemicals in the earlier chapters, is so dreaded by so many people that there ought to be warning labels on everything containing it, no matter how small the dose. In fact, there will be very little mathematics here, enough to scratch the surface but not enough to leave any permanent scars. Still, no one should be tempted to do anything against his principles, and anyone really bothered by small doses of mathematics is encouraged to skip this chapter, and go on to the Epilogue. If you want to come back and take a crack at it later, we'll still be here.

Virtually everything that has been said about risk has had an element of chance to it—the chance of getting killed, the chance of falling, the chance that a nuclear power plant will disintegrate, the chance that dandruff is caused by insufficient ice cream, and so on. Probability is involved in everything that is not certain, and, since nothing is really certain, probability is everywhere. Professional gamblers are truly among the few members of our society who understand probability well enough to use it to earn their livelihood. Amateur gamblers subsidize that livelihood. Yet probability and statistics are really not that hard to cope with, except for those who feel the urge to get deeply philosophical and dig at the roots. Then any subject becomes more difficult than it seemed at first glance. There is no need to do that to be able to use probability as a tool for the understanding of life.

This chapter is meant to do what the name implies—supply just enough knowledge about statistics to live happily ever after. It is not to be memorized but to be understood; nor is it meant to be read through like a novel. It would probably be best to read it in sequence, but even that isn't necessary. There is no right way to learn things, despite the posturings of the professional pedagogues. What works for you, works. This author has been a college professor for most of his life, and has come to believe that the end justifies the means in the teaching and learning business. (It has been alleged that W. C. Fields once said, "If the end doesn't justify the means, what the &%#$ does?") If you can trick a student into learning something he wouldn't otherwise have learned, perhaps by making him think the subject is more interesting than it really is, you will have done good, and will not have sinned.

So, dear Reader, probability and statistics are fascinating stuff. Not only are they useful in understanding risk, they are essential for understanding straw polls, lotteries and more rewarding forms of betting, and scores of problems of everyday life. It is a truism that the more you know the happier you will be, and that is doubly true of this subject. Many otherwise splendid sayings happen to be false, but "Ignorance is bliss" is a prizewinner.

Statistics and Probability

Probability and statistics are two different concepts. Probability is a measure of the likelihood or chance that something will happen, like the dice landing in a certain way. It is a number between zero and one, with zero representing the impossible and one representing certainty. There is lively combat among the professionals when the definition of probability is pushed further. The so-called frequentist school believes that the best definition is based on frequency. Imagine you've done something zillions of times; the fraction of times the event of interest has occurred is its probability. That sometimes works—if you toss a pair of honest dice a million times, you'll get a seven close to one-sixth of the time. So the probability is said to be .1667, or 1/6. (If it didn't happen that way, you'd conclude that the dice weren't honest.) But what if you want to know the probability that the Socialist Party will win the next presidential election—that is certainly not going to be held a million times. For such matters the frequentist definition fails, and the so-called subjectivist takes over. His definition is that the probability is the best judgment of the best-informed people, no more and no less. That works for dice; the best-informed people know how dice work, and therefore know the odds. Even frequentists, deep in their hearts, don't depend on a million dice throws to know the probability of a seven at a game of dice; they calculate it, just like the rest of us. Usually it doesn't matter how probability is defined—what matters is what use is made of it. Yet, several centuries after probability was invented, experts still fight about what it is.

Everyone agrees on what statistics are—things that
are measured or counted, and it hardly matters what. The
height of the emperor of China would be a statistic, if
there were such a person, as would the average height of
all Chinese. The average income of all left-handed Americans named Dave qualifies, as does the total number of
road fatalities over the Labor Day weekend. Anything
with that broad a definition sounds empty, but the interesting part begins when we try to find the relationship
between one statistic and another. Do Daves have higher
income if they are left-handed, or do left-handed people
have higher income if they are called Dave? Those are
different questions. Or neither? Or is it different in the
Southwest? Those are the sorts of questions statisticians
are called upon to answer. They might also be called upon
for the probability that rich left-handed Daves also have
beady eyes. Or they might be asked whether it is significant that the National League won the World Series six
out of the last eleven times, but the American League five
of the last seven (as of 1989). And the American League
twenty-seven of the last fifty. Probability and statistics
come together when questions like that are raised.

The word "significant" comes up frequently in statistical analysis, as it just did. It means that the observation
was probably not just luck. Whatever happened had a
low probability, based on pure chance, so something deeper
must be involved. The probability cutoff that makes something significant is in the eye of the beholder. In the biological sciences there is a kind of informal agreement that
the cutoff is at a probability of one in twenty, 0.05, so anything that has a random probability of occurrence of less
than one in twenty, but happens anyway, is considered significant. Among non-professionals in statistics the criteria

for significance vary widely—good engineers have defended probabilities everywhere from one in a thousand to one in ten as measures of significance. In the end, the interpretation of significance depends on circumstances that have nothing to do with probability. If the price of error is high, we want to be pretty sure of ourselves. Most people have adjustable standards, according to the importance of the estimate.

The term "random" was used above, and that also begs for definition. On this one even the revered Webster's Unabridged Second Edition is of little help—it uses the word "random" four times in its own definition. The newer Random House Unabridged (the name is a coincidence of low probability) does a bit better—an effort has been made to avoid using the word in its own definition, but the resulting definition is wrong. This is not said to criticize the authors of these two fine dictionaries, but to emphasize the fact that the concept is not trivial. Each of these authorities is trying to express the idea that what happened had no rhyme nor reason, but was, well, a random occurrence. It may be best to leave it at that, saying that random is defined by its opposite. The idea of randomness is a very deep one in mathematics, and does lack a precise definition. It's just as well, since we sometimes encounter apparently random occurrences that later turn out to have a clear pattern to them. It is the absence of a pattern that characterizes randomness, but the absence of a pattern is as often a function of the observer as a feature of the observed. Smart people see patterns where the rest of us don't, while charlatans, believers in the supernatural, and cloud watchers see patterns that aren't there. The late great mathematician John von Neumann said, "Anyone who considers arithmetical methods of producing random digits is, of course, in a state of sin." We'll let it go at that.

Two of the laws of probability are most important. The first is that if we want to know the probability that both of two independent events occur, and we know the probability of each alone, the joint probability is obtained by multiplying the two together. When I throw a quarter into the air, the probability that it will land heads is one-half, and the same is true if I throw a dime. If I now throw up both a dime and a quarter, the probability they will both show heads is one-fourth. The multiplication rule applies to cases in which both events need to occur. It is an "and" rule.

The second rule is an addition rule, and applies when two events are mutually exclusive, which means that they can't both happen. A coin can land heads or it can land tails, but it can't do both at the same time. A person's last name can begin with W or it can begin with G, but it can't begin with both. For such cases, the probability that one or the other will be true is the sum of the probabilities for each alone. It is an "or" rule. For example, if we toss a pair of dice, the chance that we get a four is one in twelve, a ten is also one in twelve, and a four *or* a ten, one in six. If the cases are not exclusive the situation is more complicated. These two rules are important.

Populations and Samples

In statistics it is often necessary to distinguish between a population and a sample of that population. One of the commonest statistical problems is to infer some characteristic of a population by sampling, and then to estimate the accuracy of the result. In every presidential election we are barraged by the results of polls which report that a random (note the word) sample of a thousand voters indicates that fifty million of us will elect Candidate Z on election day. Nowadays the sophisticated pollsters will add that the margin of error for this poll is $\pm x\%$, where x can be anything. They may even say that this is sampling error.

We usually think of a population as a very large group (though it needn't necessarily be large), from which we take a much smaller sample, small enough to study. We may do this to learn something about a large population, as for a presidential election. We may have a production line of widgets, and look at every hundredth (randomly, of course) to see if it meets specifications, assuming it is typical. This process of guessing the characteristics of a population through sampling is called inference. It is of course not good enough to just guess—there also has to be some way of saying how well we think we have done, and the expression of that is called the confidence interval. When a pollster says the poll is subject to sampling errors of plus or minus $x\%$ he is speaking of a confidence interval. It isn't necessary to know the formula for a confidence interval, only that it is a measure of the range within which the right answer for the population lies, with some high probability.

All this has been a matter of learning about a population through sampling. Sometimes the job is to use population statistics to learn about a sample. Then probability is more directly relevant than statistics. Suppose we know that 51.3% of the population of the United States is female (it is, because we count noses at census time every ten years), and want to organize a dance by picking twenty people at random, regardless of sex. What is the probability that we will have ten paired couples? The answer is 0.175, a bit better than a chance in six, and just a shade worse than if the population were evenly split between males and females. If the experiment were repeated for people over a hundred years old, where females outnumber males by more than five to one, the odds would be much worse, a chance in three thousand. The population ratio tells us how to calculate probabilities for the sample.

A population is a reservoir of some characteristic, and a sample a random selection from that reservoir. Randomness and how to assure it are constant problems of sampling. One of the most celebrated failures of a presidential poll (a failure that caused the death of the magazine that sponsored it, and gave polls in general a bad name for many years) is blamed on the fact that the polling was done by telephone at a time when telephones weren't as universal as they now are. This selected a non-random sample of the population, the relatively affluent, which biased the results of the poll. It takes great care and skill to randomize a sample so that it properly reflects the population from which it is drawn.

Hypothesis Testing

Hypothesis testing is one of the harder jobs posed to a statistician, and one of the most important in the risk business. It comes up in determining whether a given influence causes or cures a disease, or whether a new scientific theory is "true," and with what confidence we believe the conclusion. In the case of a scientific theory, there is a more pragmatic operational criterion—it is true when most reputable scientists believe it to be true.

Suppose ten people have a given disease, called D, and an experimental drug, called C, has been proposed as a cure. It should be tested (after the mandatory testing for side effects on experimental animals), so the sick patients are divided into two groups of five each, and the drug is administered to one of the groups. To be sure that the test is fair a placebo (something with no medical effect, other than psychological) is given to the others, the control group, and no one is told which patient got which. If neither the doctors nor the patients are told who got what, it is called a double-blind test, and is considered the only sure way to avoid bias. There are sealed records, so we can find out later. There is always an ethical question in such matters, the search for knowledge that may benefit everyone for a long time competing against the possibility that some patients may be denied an efficacious drug. There is no easy universal answer to that question.

The hypothesis being tested is that the drug helps cure the disease, as compared to the so-called null hypothesis that it doesn't. The test is construed as a test of the null hypothesis. Suppose that three of the group who got C recovered, and only two of the other group, what have we learned? Of course we have learned that D is not always a fatal disease, because even two of the untreated group recovered, but only a very few diseases are uniformly fatal. The question is whether the experimental drug helped tip the odds in favor of recovery. All of medicine works on statistics, tipping the odds, because although few diseases are always fatal, few cures are always effective. How are the test results interpreted?

The standard procedure for this type of problem is to make a kind of table to portray the situation. It is called a contingency table, and is shown on the next page.

The challenge is to decide whether a table of this sort could have resulted from random chance (testing the null

Treatment	Cured	Died
Treated	3	2
Untreated	2	3

The results of treatment

hypothesis). Of course it could have, but with what probability? It is evident that the treated patients did somewhat better than the untreated ones, but what is the chance of that if the treatment was in fact valueless? Even worse, what if the treatment was slightly harmful?

The test usually used under these conditions is called the Fisher Exact Test. (This author doesn't like it, for reasons that will appear later.) It consists of asking for the probability that the distribution in the table, or one with even better results, might arise if there were really no difference between the treated and the untreated cases. The answer given by the test for this case is exactly even money—results at least this favorable will occur half the time by chance alone—so one should put no credence whatever in the apparent evidence that the treatment has done some good. The rule of thumb in the biological community, as mentioned before, is that the probability that the table (or an even more favorable outcome) could have come about by chance, *calculated according to these specific rules*, must be less than 0.05, one chance in twenty. By that standard, the treatment would not have been credible even if the numbers above were 4 and 1, but would have been if all the treated patients survived and all the untreated ones died. This contradicts most people's instincts, which tell them that a treatment that cures 80%

of the patients is pretty good, when compared with only 20% survival without treatment. It might look good, but still wouldn't pass the Fisher test for significance, using the standard criterion.

We've gone through this particular test in some detail, but there are many other versions of hypothesis testing, or significance testing, depending on the form and type of the data. A baseball enthusiast might look back at the last fifty years of baseball history, to ask whether the total number of runs scored in the major leagues is related to the average summer temperature in the Midwest. He would collect the data and probably do what is called a linear regression, to see whether the number of runs seems to show a trend of increasing (or decreasing) with the temperature. There is a significance test for that kind of problem too.

Closer to our main subject, he might ask whether air pollution contributes to mortality. He would then collect the mortality statistics on a dozen or so cities, along with the various measures of air quality, to test for trends and for significance level. The exact procedure goes beyond our self-imposed rule of "just enough statistics," but it is also a linear regression analysis, with more variables. This one has been done by several analysts, and leads to the conclusions we used in Chapter 16. Nowadays statisticians have access to computer programs that will do this kind of assessment for them, so even they no longer have to understand what they are doing.

Sometimes there are no cut-and-dried statistical tests available for a problem, and statisticians may differ on whether the hypothesis is true or if the data are significant. The formaldehyde data in Chapter 12 are in that category, probably insignificant at low levels, but not so much so as to deter the regulatory agencies. Where the data are adequate, the standard tests of significance are fine, but where they are inadequate, interpretation of the data is an art form.

Parameter Estimation

For most situations to which we apply probabilistic reasoning, there will be what is called a probability distribution. That means that there are a number of possible events or outcomes of a test, or survey, or whatever, and that each has a certain probability attached to it. It is customary to plot a graph to display such information. For instance, if I throw a handful of four coins in the air, the probability that they land showing four heads is one-sixteenth, the chance of three heads and a tail is one-fourth, of two heads and two tails three-eighths, and so on. That is a probability distribution, a display of a collection of possible events and their probabilities. There are different more or less standard distributions, appropriate to specific situations.

Once the specific statistical laws for a situation are understood, the characteristics of the probability distribution are known and we only want to know the actual numbers. Suppose we have an unbalanced coin for which the probability of tails is higher than the probability of heads (we'll see a real-life example in a moment); we might then want to know the actual probability of tails. Once we know that we can predict the probability of, say, eight tails in ten tries, or seven or six or whatever. The statistical law that governs this sort of thing is called a binomial distribution (next section), but it can't be used unless we know the underlying probability of a tail in a single try. To find that out we would have to find the average fraction of tails in many tries, the so-called parameter of the distribution. We could then calculate how likely it is that a certain number of tails will appear in a certain number of tries, using the binomial distribution. In other words, we will learn about the fluctuations around the average, just by measuring the average. This can only happen because we know the statistical rule. So we measure the average for a large number of tries and estimate the small-sample probabilities from the large sample.

The average is a simple parameter, but there are other interesting parameters of a probability distribution. The average IQ of the population is 100, because that is the ultimate definition of IQ. (We can't all be above average, which is depressing, but we also can't all be below average, which is reassuring.) Of course, some people are above average and some are below, and there are people who are far above or far below—that is everyday experience, without the benefit of IQ tests. It is known from many decades of testing, and from many types of tests, that the distribution is what is called a normal distribution (next section), an unfortunate name for a specific mathematical form. What is important for society is the *spread* of the distribution, which tells us how many gifted children and adults there are likely to be, and also how many will need special help. The spread, or width, is therefore another parameter that can be estimated by measurement. That parameter can only be found by testing many people, to see how their IQs vary around the average. About two-thirds of us are within fifteen points of the average all-American IQ.

There are many things in life that are well described by the normal distribution—birth weights of babies, heights of adults, running speeds of randomly selected people, etc. It is ubiquitous in statistics. There is a powerful theorem, impressively called the central limit theorem, which asserts that nearly all probability distributions are similar to the normal distribution, if only the number of cases involved is large enough. That will be obvious for some of the specific distributions that follow.

Standard Distributions

Some probability problems are easy to solve mathematically. A problem is easy if you don't have to know many details in advance to compute the probabilities. The classic example is flipping a coin. The reasoning goes like this: heads and tails are mutually exclusive—the coin can't do both at the same time—so whatever may be the probabilities of a head or a tail separately, the combined probability that one or the other will happen is the sum of the two. That is the "or" rule. But there is no reason for the coin to prefer heads to tails, so the two probabilities are bound to be equal. The sum of the two probabilities, heads or tails, has to add up to a probability of one, since it is a sure thing that the coin will land somehow—no one can throw a coin into orbit. If two equal numbers add up to one, they must each be a half. We therefore know, without flipping a single coin, that the probability of heads is a half, 0.5.

This fact is so instinctive that we never think the logic through. Exactly the same argument tells us that the chance of a single die landing on any given side is one-sixth, and when we multiply the probabilities for two independent dice (the "and" rule) we find that the probability of snake-eyes (two ones) is one in thirty-six. We can learn the same way that the probability that we are dealt a royal flush in poker is one in 649,740. We don't learn that by playing through a few million hands to see how often a royal flush appears, we learn it by calculation. Whenever there is a situation in which all the different possibilities are interchangeable and exclusive, so the probabilities add up to one, we can easily and confidently work out the probability of each.

But it is important that we be certain that the different options really are the same. One of the articles of faith for the average American is that an honestly thrown coin has an even probability of landing heads or tails. That's the origin of the term "tossup." It is surely true if the coin is thrown high enough to "randomize" the throw. (In a casino the dice have to be thrown against a board to achieve the same randomization, lest devoted gamblers practice for years to get some control over the outcome.) Nearly all people believe that the same degree of randomization can be obtained by spinning the coin on a table— the heads and tails go around so fast that it's hard to see how one can be preferred over the other.

Yet it isn't true for some coins, like an American cent. If a cent is well spun on a smooth table, it will come down tails about 70% of the time, because there really is a small difference between the head side and the tail side. The reason the slight difference matters is a fairly complicated feature of the physics of rotation, but matter it does, and this author has confounded many a friend with the demonstration, as he himself was once confounded. (If you look at the top of the rotating cent, you will see that the axis of rotation isn't exactly vertical, and that a little imaginary circle seems to be formed at the top. It is this little circle that biases the way the coin lands.) To achieve the same effect with dice it is necessary to load them (and that is illegal), but a cent is minted in loaded form. Beware of honest-looking people who spin coins rather than flipping them.

Many probability distributions can be calculated with these methods, and it turns out that a small number of standard distributions is sufficient to deal with nearly all cases.

Recall that a probability distribution is just a portrayal of the probabilities for different outcomes of a test. Sometimes the absolute probability that something will happen is not as interesting as the relative probabilities of the various alternatives. A coach in the National Basketball Association might find it useful to know what fraction of people are over seven feet tall, compared to those between six and a half and seven. From that he might learn the probability that he'd see a possible seven-foot recruit going by if he stood on a street corner in Pittsburgh for a day. The birth weights of babies follow a normal distribution pretty closely. If we knew its parameters, as we do, we could compute the probability of an eight-pound baby, something that ought to interest any prospective mother. (A little more than one chance in ten, according to some old data.) Any of these is a probability distribution, which can be displayed on a graph of probability plotted vertically against the other variable of interest horizontally, whether height or birth weight or IQ or degree of drunkenness (for the probability of a crash).

The standard distributions come up so often that they have names, and each is applicable to a certain type of situation. The most important, in terms of how often they come up in everyday life, are called the Poisson distribution, the normal distribution (often called Gaussian), and the binomial distribution. Most of the subjects in the book are loyal to one or another of these forms of statistical behavior.

Poisson was a French mathematician, and Gauss a

German mathematician. Normal is just an English word. Binomial is a mathematical term, and the binomial distribution is sometimes called the Bernoulli distribution, after a member of a famous family of Swiss mathematicians. Science and mathematics are international.

The Poisson distribution is applicable to a random counting process, in which each event is independent of all the others. We might count the number of phone calls that come into a switchboard in each five-minute interval, the number of cars passing a point on an uncrowded freeway every minute, the number of cosmic-ray particles raining down on the author's head each second that this is being typed, or the number of cases of cancer appearing in an exposed population group. For each of these examples there will be some long-term average, the parameter for the distribution. In the car case, we might count the number of cars that pass in an hour, and use one-sixtieth of that as the average for each minute. Then, for each of the measured minutes, the number that pass may be more or less than the average—just random fluctuations—and the Poisson distribution gives the relative probabilities for each. (For cosmic rays on the author's head, the long-term average is about five per second, at sea level.)

The top figure on the next page shows the Poisson distribution when the average count is three, and we see that there is a good chance that none at all will pass in any given minute (for the car case), while there may instead be six or seven or even more. The most probable numbers are two and three (which have equal probabilities in this case), but in fact any number is possible. Of course the very large numbers are highly unlikely if only three pass in the average minute—we see that from the way the bars get smaller as we move to larger numbers. (When they are

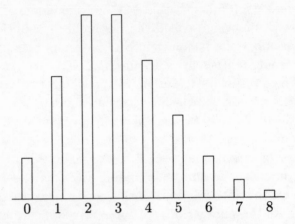

Poisson distribution for an average of three counts

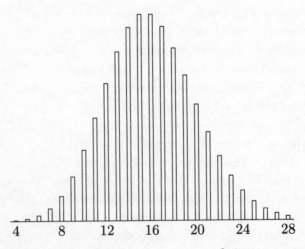

Poisson distribution for an average of sixteen counts

bumper to bumper, all bets are off, because they are no longer independent.) Still, it is a very broad distribution, with lots of opportunities to vary from the average, and the fluctuations are precisely known once we know that the average count is three.

The bottom figure also shows the Poisson distribution, but this time with an average count of sixteen. (There is a distribution for every possible value of the average, even when it isn't a whole number.) Note the changes. There is now a much more pronounced peak around the value of the average, and the chance that none will pass is now negligible. The most important single fact about the Poisson distribution is that its half-width, which is just what the name implies—half the apparent width of the distribution—is about equal to the square root of the average value. This is what was referred to in the text as the square-root-of-N rule. Thus, in the figure for an average of sixteen, we would expect a range of plus or minus four to be reasonable. If we look at the figure, and find the vertical bars corresponding to twelve and twenty, four below and four above the average of sixteen, we can see that the distance between them is a reasonable rough measure of the width of the distribution. (Each vertical bar measures the probability of that outcome.)

This will always work for a Poisson distribution, and is the basis for the little note that appears at the bottom of most public opinion polls, saying that the margin of error for the poll is such and such. This will always be approximately the square root of the number of people polled, divided by the total number. (There is a small refinement having to do with the fact that a vote given to one candidate is taken from the other, so they are not entirely independent.) That's why you can't interview a

few hundred people and find out to within a couple of percent how the election is going to turn out. If you interview
two hundred people in a close election, you expect about
a hundred votes for each candidate, so the random fluctuations will be about ten votes for each, from sample to
sample. Therefore no small sample will provide enough information to call the election closely. These are statistical
issues, independent of all other polling problems.

The next important probability distribution, the most
important of all, is the normal distribution, shown in the
figure. It is universal in statistics, and we have already
said that the central limit theorem tells us that all other
distributions look like the normal distribution when the
numbers are large. This is already happening to the Poisson distribution when the average is only sixteen, and the
resemblance would become more and more striking if we
were to go to a hundred or a thousand average counts.

We usually expect the normal distribution when we
are measuring something, rather than counting it. We
have said that it describes the variations in birth weight of
babies, IQ of children and adults, heights (but not weights)
of adults, and things of that sort. Since the Poisson is so
like the normal when the counts are large, most statisticians will use the normal distribution even in that case,
just for convenience.

One essential difference between the two is that the
"width" of the Poisson distribution is entirely determined
by the average number of counts expected, via the square-
root-of-N rule, but the width of the normal distribution
has to be separately measured, and is the most important parameter of the distribution. There is a name for
this measure of the half-width—it is called the standard
deviation. (There is a standard deviation for nearly all

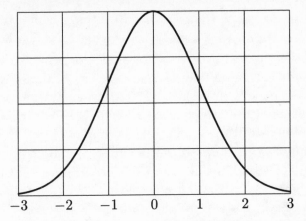

-3   -2   -1   0   1   2   3

Normal distribution for a standard deviation of one unit

distributions, and it measures their width.) The figure above shows a normal distribution for which the standard deviation is one square. We have also deliberately plotted the second Poisson curve so that the standard deviation of four has the same width as one square on the normal curve. It is apparent from looking at them that it is not too bad an approximation to use the normal as a substitute for the Poisson, as could have been guessed from the central limit theorem.

The normal curve illustrates a point that came up earlier in the book, that random fluctuations around the average (or mean) are to be expected, but that fluctuations of more than two or three standard deviations are very improbable. The curve is practically down to zero by the time we are three standard deviations from the center. If we know the standard deviation, either because it has been measured or estimated, or because we are using the normal distribution as an approximation to a Poisson (so we can take a square root to find the standard deviation), then we know the likelihood of any deviations from the mean.

Converting that comment to a statement about significance, it means that a fluctuation of one or two standard deviations is to be expected from random chance, but a measurement more than that far away from the expected value is unlikely to be just a statistical aberration, and is significant. (The sampling error quoted by most pollsters corresponds to two standard deviations.)

Finally, we come to the binomial distribution. That applies to situations like the coin-tossing case, the champion batter we used as an example in Chapter 5, and other cases in which we know the underlying probability that something will happen on each try, but want to know the probability that it will happen a specified number of times in a sample of a specified number of tries. In other words, it tells us about the fluctuations in the count for a finite number of samples, when we know the long-term or population average.

The classic case is tossing coins. Assuming honest coins (it is good to be suspicious when gambling) we know that the chance of heads is 0.5, a tossup. If we toss ten coins at once, then we know that the *average* number of heads will be five. Suppose, however, that we ask a different question: what is the probability that we see *exactly* five heads in the ten coins? That is an entirely different question, because the answer to the first includes the fact that we will have six as often as four, seven as often as three, etc. The answer to such questions is given by the binomial distribution, and in this particular case, even though five is the average answer, and even the most common result, the probability of exactly five heads is only 0.246, less than a chance in four. (Try it—this is an easy experiment. Of course beware, because even experiments of this sort are subject to statistical fluctuations, and you might have to do the experiment twenty or thirty times for the average to show up clearly. That would involve a few hundred coin tosses, but a few minutes is a reasonable expenditure of time for learning something. It's a bargain.) This was an easy case because the underlying probability was 0.5 for each coin, but it is only slightly more complicated (the formula, not the concept) when the probability is not 0.5. The powerful batter we used in Chapter 5 had a batting average of .300, which is his underlying probability of getting a hit each time at bat. (We are ignoring baseball-specific complications in counting times at bat, like walks, errors, sacrifice bunts and flies, and the like.) The final figure, on the next page, is for this case, and shows the probability that he will get a certain number of hits in a random selection of fifty times at bat. Because fifty is a reasonably large number, the distribution is not very different from that shown for the Poisson distribution for an average of sixteen (here our average is fifteen), in conformity with the general principle that nearly all distributions look alike, and look like the normal distribution, for sufficiently large numbers. (Though we've called this the central limit theorem, many statisticians simply call it the law of large numbers.)

The reader will have noticed by now that there is a family resemblance among these distributions. They each have an average (or mean), which is 3, 16, 0, and 15 for the four cases in our figures, and they each have a half-width (standard deviation), which is 1.73, 4, 1, and 3.24 for our four cases, respectively. These basic features are common to all (to be completely honest, nearly all) probability distributions, so that recognition of these is sufficient to understand the statistics in nearly all cases. For Poisson-type cases, there is even the advantage that the standard de-

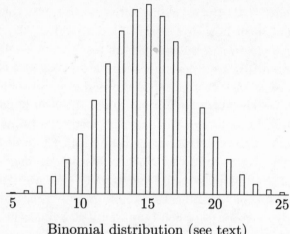

Binomial distribution (see text)

viation is equal to the square root of the mean, while for a normal distribution it has to be found some other way, and a binomial follows a small modification of the Poisson rule. If you know these things, and appreciate the importance of the standard deviation as a measure of the width of the distribution, you know enough.

Bayesian Statistics

This section goes beyond what you need to know to deal reasonably with risk, and is purely for deeper understanding.

Thomas Bayes published his work in 1763 (actually his friends published it for him posthumously), and it still causes tempers to flare among statisticians. The problem is the following: though we've discussed several ways to use population data to infer sample probabilities, there is no clean way to do the reverse. We alluded to this when we spoke of parameter estimation, and it is involved in hypothesis testing, but it is not a solved problem in classical statistics. It reverses the order of normal probabilistic reasoning.

As an example to which most can relate, suppose you get into a game of dice with friends (dice are the oldest gaming objects known), using the usual rules of craps—if you don't know them we are not going to teach gambling here—and wonder what the odds are of seeing a run of four consecutive naturals (seven or eleven). An easy calculation shows that there are two ways of getting an eleven and six ways of getting a seven, out of thirty-six distinct ways the dice can land. The probability of a natural is therefore eight out of thirty-six, or 0.222.... That is the normal use of probability, and also tells us that the chance of two naturals in a row is approximately 0.0494 (the "and" rule), of three 0.011, etc. The odds against four in a row are more than four hundred to one. So far so good.

Your friend the shooter tosses the dice and gets a natural, so you congratulate him. He does it again, and you congratulate him again. The third time you begin to wonder (the odds are nearly a hundred to one), and by the fifth you are convinced that your (former?) friend was cheating. You have just made use of Bayesian statistics. Having calculated that the odds of even four in a row challenged your credulity, and the odds of five happening by chance were 1844 to 1, you concluded that you should question your original premise that your friend and the dice were honest.

Bayesian statistics involves learning something about the assumptions by looking at the results. Though heartily disliked by most classical statisticians, it is used instinctively by everyone, as in the example. In its formal structure, as first stated by Bayes more than two hundred years ago, it provides a quantitative way to evaluate the probabilities of different assumptions, given the data.

In science, for example, there are often competing hypotheses for the explanation of some natural phenomenon, and you may even be able to find bets among scientists about which is most likely to be right. (Three to one on general relativity, and that sort of thing.) Then, after an experiment has been performed to "test" the various hypotheses, the odds will have changed, just as the odds of a particular team winning the World Series change after the first game has been played. In the case of general relativity, probably the only theory in the modern history of physics to be formulated and proposed without any experimental evidence that needed explanation, it took three famous experiments before it was generally accepted.

Going backward into the unknown, using the observations, is what characterizes Bayesian statistics—it can't be done in classical statistics. When we spoke of the Fisher Exact Test earlier, we really wanted to know whether the treatment could cure the disease. Instead we assumed for the purpose of the argument that it could not, and calculated the probability that the observation *could* have come about by chance. We made an assumption and calculated the outcome, though we already knew the outcome, and were interested in the validity of the assumption. This can only be done in Bayesian statistics.

It is because of this lack that the interpretation of the Fisher Exact Test (which we are just using as a whipping boy here) is so arbitrary. If the probability that the observation could have come about by chance is less than 0.05, the data are considered significant, otherwise not. There is no rationale whatever for the arbitrary number 0.05, yet it is used as a criterion for significance of many tests that have regulatory consequences, like whether a given chemical causes cancer in rats.

If we apply this criterion to the dice example above, we should congratulate our friend on his luck if he tosses one natural, but conclude that the dice are loaded if he does two in a row (for which the probability is .0494, less than .05), and pursue the consequences. In a real case of that sort, we might use that uncommon quality called common sense, and take into account our long friendship, the intrinsic trustworthiness of our friend, and the difficulty of finding a pair of loaded dice. In short, we might behave sensibly, though subjectively. Subjective is not a dirty word, even in statistics.

Bayesian statistics has been controversial because it inevitably involves a certain amount of subjectivism. That deprives it of the neat formal structure enjoyed by the classical form, and deprives the practitioner of his sense of orderliness and security. People don't react well to such deprivation. For the reader, who presumably has no aspirations to become a professional statistician, the only lesson of this section is that it would be wise to keep an eye on whether people are using data to test hypotheses or using hypotheses to predict data. They are different activities, and you will find that they are often confused. Much of life consists of the former, and much of statistics consists of the latter.

20

Epilogue: What Does It all Mean?

After all that, has anything useful been said? Many different sources of technological risk have been covered, and it should have become clear that they have some characteristics in common. By far the most important is that they are all small, and pose almost imperceptible threats to most of us—the vast majority of us are doomed to die of far more mundane causes than strange chemicals, pesticides, or radiation. Even vehicular travel, the largest single source of technological risk, and a real killer, is responsible for "only" one death out of forty in the United States. Apart from that, technological risk simply doesn't find its way into the standard mortality tables.

Nonetheless it is real. The risk is there, sometimes lurking as a risk-in-waiting, with little threat now but the potential for real problems later. There is real risk in nuclear power, just as there is real risk in coal power. The latter kills now; the former may do so later. Acid rain is here now, while the greenhouse effect may not yet be quite upon us. These are matters of great importance, but invisible in terms of current mortality. For some of them, like the greenhouse effect, the potential damage is devastating, while for others, like nuclear accidents, the risk is limited, but imaginations are not. For still others, like the risk posed by a high-level nuclear waste repository, there is essentially nothing outside the imaginations of the gullible.

For the technologies that carry risk there are also benefits. Anyone who tried to take away people's cars because they are risky would become very unpopular very quickly. Even those who try to make driving safer are unpopular if it causes inconvenience. The anti-nuclear groups who want to shut down the 20 percent of our electricity that is supplied by nuclear power had better hope they don't succeed, because the population might then turn on them as the lights and air conditioning went out. Similarly for the crusaders against some of the truly negligible risks from the chemical residues in our food. If they really threatened the supply of food, their support might evaporate. It is easy to moralize and behave as if there are no benefits, as long as our comfort and our normal way of life aren't threatened. In short, if it is some invisible or distant ox that is being gored.

It cannot be a matter of reasonable dispute that technology has had a dramatic positive influence on the safety, health, and longevity of our population. Our life expectancies have not gone up by more than twenty years since 1920, and doubled in the last hundred-odd years, just because the stars are in the right places. It is because technology has given us safer and easier ways to do the things we do, has provided energy to replace much of our manual labor, has provided through science ways to make nature an ally in our activities, has endowed us with an enor-

mously productive agricultural system, and has supplied
the means for both prevention and cure of many of the
most life-threatening scourges of mankind. To believe that
we can simply reject technology and go back to a simple,
wonderful, safe life is just plain nutty.

But there is risk. In a more perfect world, we would
make reasonable judgments about the risks and other costs
of specific technologies, and match them against the benefits. We do something like that when we buy a banana
split. Yet for risk we are hardly ever willing to do it openly,
preferring to make our decisions through such complex
mechanisms that we don't know what we are trading for
what, or who is responsible for the decision. We seem to
like it that way.

Even if we tried to make the tradeoff as well as we
could, we would fail in most cases, because there is almost
always uncertainty in the assessment of a risk, and the
benefits are at least as hard to quantify. We have recently
had a national scare about a particular chemical residue on
apples—school boards across the country took apples out
of school lunches, and grocers cleared their shelves. Little
does it matter that the threat was greatly exaggerated,
propagated by an organization that earns its livelihood
through such alarms. The scare was real enough, and even
the risk was real, though small. It didn't have much impact
on the students' health, because there are other fruits to
be had, but it would have been damaging if that were not
true. At such a time, people who say that we ought to
balance the minimal threat of the residue against the fact
that fruit is healthful, are not popular, and are certainly
ineffective. Kipling wrote a famous poem whose first lines
are, "If you can keep your head when all about you, Are
losing theirs and blaming it on you ...." It isn't easy,

and Kipling forgot to mention that the best way to keep your head on your shoulders under such conditions is to keep your mouth shut. We come from a long tradition of beheading the messenger who brings unwelcome news.

To make any of these assessments responsibly some effort at a quantitative treatment is absolutely essential. It isn't enough to say that such-and-such a chemical additive causes cancer, or that flying is dangerous; that approach leads to paralysis. It matters just *how* dangerous the chemical may be, or just *how* likely is an accident. It's impossible to eliminate risk entirely from our lives (and wouldn't even be desirable if it were possible), so the need to know what we are up against is paramount. We all (well, almost all) learn in late childhood to do that with our personal budgets—we outgrow (or manage to suppress) the urge to have every toy in sight, and learn to ask whether a particular acquisition is worth the cost. It is no different with risk avoidance, it is just less familiar.

We also seem to be losing our ability to recognize that there are subjects on which reasonable people can disagree, with neither side villainous. In 1989 *The New York Times* ran an article entitled "In Science, Error Isn't Fraud." The most common popular response to any unpleasantness is that someone has inflicted it upon us, for unimaginably evil reasons, or perhaps only to make a profit at our expense. Our courtroom drama, both real and fictitious, is predicated on the idea that someone is to blame for anything bad, and that it didn't have to happen.

It is no surprise that we think that way. From the exposure we and our children have, to the television programs, to the books on newsstands, to the most popular movies, it is hard to see where we can learn that bad things can sometimes happen without an evil force behind them.

There is always a person wearing a black hat—that is the stuff of which fiction is made. It is also much easier on the intellect to believe in witches and villains than to confront problems that require judgment. The much-maligned turn to the negative in our political campaigns is just another recognition of this need.

So we live with a system in which there are technological risks and benefits, each of which is hard to quantify. Much of our leadership in these matters is filtered through the media, with a natural tendency to exaggerate risks and to search out villainy (in the apple panic, this author saw no newspaper or magazine article or television program that even mentioned that apples are healthful). Too much spleen has doubtless been vented in this book against those who exploit our weakness in these matters, but they are (often with the best of intentions) making this a less safe world for the human race. The fact that, as a population, we are so uncomfortable with and ignorant of science and technology, and innumerate to boot, seems to make us irresistible targets. Beware of pseudo-experts with a mission and a grudge, especially if they are lawyers pretending to be scientists.

In the end, there is no substitute for education. If any deny that there are risks in technology, they are just as wrong as those who think that a simpler world would be safer. The former contribute to the complacency that breeds the accidents we do have, and they require occasional shock treatment. For the latter, it wouldn't be compassionate to begrudge them their dream, but they don't have to foist it upon the rest of us. As in all real-life situations, there are conflicting values that need to be reconciled, someone's ox does need to be gored, and it matters whose. The only way to deal effectively with these genuinely deep issues is to understand them—we can't always count on luck. This book is a modest effort to help. It would be even better if you enjoyed reading it.

Index